DATE DUE

			PRINTED IN U.S.A.

PHILIP L. FRADKIN

FALLOUT

—— AN ——

AMERICAN

NUCLEAR

TRAGEDY

The University of Arizona Press / Tucson

Second printing 1989

THE UNIVERSITY OF ARIZONA PRESS

Copyright © 1989 by Philip L. Fradkin
All Rights Reserved

This book was set in 10/13 Linotron 202 New Caledonia.
Designed by Kaelin Chappell
Manufactured in the United States of America
♾This book is printed on acid-free, archival-quality paper.

93 92 91 90 89   5 4 3 2

Library of Congress Cataloging-in-Publication Data

Fradkin, Philip L.
    Fallout: an American nuclear tragedy/Philip L. Fradkin.
        p. cm.
    Bibliography: p.
    Includes index.
    1. Nuclear weapons—Nevada—Testing.   2. Nuclear weapons testing
victims—Nevada.   3. Nuclear weapons testing victims—Utah.
4. Radioactive fallout—Nevada—Physiological effect.
5. Radioactive fallout—Utah—Physiological effect.
6. Liability for nuclear damages—Nevada.
7. Liability for nuclear damages—Utah.   I. Title.
U264.F72 1989
363.1′79—dc 19   88-27813
                                                                          CIP
Cloth ISBN 0-8165-1086-5 (alk. paper)
Paper ISBN 0-8165-1143-8 (alk. paper)
British Library Cataloguing in Publication Data are available.

For my mother,
Elvira Kush Fradkin,
who wrote some sixty years ago,
"Peace rests with us."

CONTENTS

ACKNOWLEDGMENTS

THIS WAS a difficult book to research, write, and get published.

There was a vast amount of primary source material that had to be gathered and digested. I sat through the lengthy Allen trial in Salt Lake City and took notes of the proceedings. During a brief recess, or after the day's session, I talked to the lawyers and witnesses in the hallway or in a nearby coffee shop. I was allowed to sift through the boxes of exhibits, notes, and miscellaneous documents that the government lawyers had accumulated and then stashed in an office on the top floor of the Salt Lake City federal courthouse at the conclusion of the trial. I also obtained a copy of the trial transcript. United States attorneys Ralph Johnson and Henry Gill were particularly helpful, as were the plaintiffs' lawyers. Stewart Udall, J. MacArthur Wright, Dale Haralson, and Ralph Hunsaker gave me access to their files.

After I absorbed all the legal materials, I decided to use the pretrial maneuverings and the trial as the framework for the narrative. In itself the civil trial, whose prominence would be short-lived, lacked coherence and drama. Complex scientific explanations by people with advanced degrees do not lend themselves to Perry Mason–like confrontations. Use of the legal motif, but not the straightjacket of legal procedure, enabled me to tell the story in a manner that suggested crime and punishment (or the lack of it). There is a sampling of events, the discovery of the alleged crime, the failure of redress outside the courts, the sweep of historical events, the victims, the evidence, and the judgments, which were comments as much upon the events as upon the adequacy of the institutions. The legal proceedings did furnish some of the raw materials and the structural device for this approach to the story of government malfeasance.

Following the trial, I visited some of the plaintiffs and their relatives in their homes in town or on their ranches in Arizona, Utah, and Nevada. Invariably they made me feel at home, although the subject of our conversations was quite trying for them. I hope that I have repaid their trust by not trading to an unnecessary extent upon their grief. One editor said that an earlier version of this book was not wet enough. I think he was referring to tears. It was that type of approach that I sought to avoid.

During my wanderings a number of friends welcomed me in their homes and offered me a temporary refuge from the dread that the subject of nuclear fallout and its consequences evoked. They included Milton Voigt and Twinkle Chisholm in Salt Lake City, Karen and Alex Riley in Tucson, Kate Lyman in Taos, and Bill and Lucky Marmon in Washington, D.C. Believe me, their warmth was greatly appreciated.

I then spent a block of time at the Coordination and Information Center in Las Vegas, where the Department of Energy (DOE) has amassed tens of thousands of documents, a veritable treasure trove of nuclear test information. Chris West of DOE's public information staff gave me a one-day tour of the test site between my bouts of document reading. Other obligatory stops on the research trail were the Bancroft Library at the University of California in Berkeley and the library at the Los Alamos National Laboratory, or at least the unclassified section of it. Back East I searched the old Atomic Energy Commission (AEC) archives in Germantown, Maryland; Harold Knapp's nearby basement repository; the Library of Congress; and the National Archives. I also made intermediate stops at the Truman library in Independence, Missouri, and the Eisenhower library in nearby Abilene, Kansas. Within the fallout region, I perused the local history sections of the Utah Historical Society library in Salt

Lake City and the public libraries in Cedar City, St. George, Kanab, and Las Vegas. To all those custodians of written materials who aided me along the way I give my heartfelt thanks. Although many of the documents related to national security matters and thus were sensitive, I cannot recall anyone who stonewalled my efforts in an obvious manner.

I read approximately sixty books that related to the subject in varying degrees and formally interviewed or informally talked to about 150 persons, some of whom were experts in their fields of science and law. I have no background in science or the technicalities of the law. However, during a professional writing career of some thirty years I have found that enough specialized knowledge can be absorbed from others in order to pass the pertinent information on to readers. I thank those experts for their patience.

To guard against inaccuracies and give me feedback on my interpretation of the events, a number of persons who were involved in the nuclear tests and the Allen suit read portions of earlier versions of the manuscript. They were Dale Haralson, Ralph Hunsaker, Stewart Udall, Dan Bushnell, Harold Knapp, Joseph Lyon, John Gofman, Norris Bradbury, Gordon Dunning, and Judge Bruce Jenkins—after he filed his opinion. Any remaining inaccuracies in the text are my responsibility, as is the interpretation of the events.

When I ran up against difficulties with my first editor, I went to four friends whose intelligence and incisiveness I respect. Michael and Connie Mery, Richard B. Lyttle, and Genevieve Atwood read an earlier version of the manuscript and gave me invaluable encouragement and advice. In particular, Michael and Rick were constant boosters and promoters of this project. I also sought the advice of Carl Brandt, a New York City literary agent who did not represent me but who was quite generous with his valuable time. After reading the manuscript he made clear what I was beginning to suspect: that this book was less about radioactive fallout and cancer than about government betrayal of its citizens. Nola Lewis typed two earlier versions of the manuscript before I purchased a personal computer and printer and put her out of business, at least in that respect. Near the end, Kathryn L. MacKay helped me with last-minute research in Salt Lake City; and Dianne Brent gave me personal support, which I desperately needed because this was not a happy project.

As I sat at the trial, traveled the Southwest, and then wrote at home, I took on the symptoms of the cancers that I heard others describe. It was a frightening experience. I felt weak and nauseous at times and exhibited other symptoms of that most dreaded disease. One day when I was writing a passage about Lenn McKinney, I began to bleed spontaneously from my nose just as McKinney, who was soon to learn that he had leukemia, had bled uncontrollably as he drove his truck home from Phoenix.

I knew that one does not catch leukemia like the flu, by simply being in a place, and anyway, there is a latency period. But my mind could not overcome the stabbing fear. There seemed to be an order of succession among those who had become deeply involved in this subject: Paul Jacobs, Dale Haralson, Gordon Dunning, and then Philip Fradkin? I kept hearing that litany of successive cancers.

This fear, coupled with the nature of the subject, resulted in alternating periods of sadness, anger, hopelessness, and despair— something akin, I thought, to what the true victims had experienced. These were feelings that I had to stifle as I wrote because I envisioned a documentary, not a polemic. I was somewhat relieved when I related my concern to a doctor, and he explained that most first-year medical students took on the symptoms of the diseases that they were studying. Thank you, Dr. Michael Witt.

This book had a circuitous publishing history, and this added greatly to my malaise. It started out in 1982 as a two-part series for the *New Yorker* magazine and was then to be published as a book by Alfred A. Knopf, Inc. Neither William Shawn of the *New Yorker* nor Ashbel Green of Knopf liked what I had produced. I thought it had value. We parted company. Subsequently, a number of other East Coast publishers rejected the manuscript, including one editor who mistook it for a novel. I was angry and desolate, so I put the book aside in early 1985, took a trip to the Sonoran Desert to cleanse myself, returned home, and started work on another project.

A literate friend, Gay Robbins, who owns Robbins Book Shop in nearby Petaluma, California, kept pushing me to do something with this book. I finally assented in an unguarded moment, and she telephoned John Little, a sales representative for various publishers in the West. Little called the University of Arizona Press, which had previously reprinted a book of mine in a trade paperback edition. The press was immediately receptive to this book, and the manuscript successfully survived the readings of two historians and the judgment of one committee. I thank all those who participated in that process for rescuing the manuscript from the oblivion of my office closet. Looking back, if it had not been for Gay and John, perhaps there would have been no book. I certainly hope, for their sakes, that they sell a few copies.

PHILIP L. FRADKIN
Inverness, California
April 1988

FALLOUT

PROLOGUE

THE CRIME

EVERYTHING that could go wrong with Shot Harry went wrong. Harry was jinxed, as the newspapers put it.[1] At the time it seemed that the Atomic Energy Commission could only be blamed for poor judgment and faulty planning, compounded by bad luck. Years later, when it became known that a subtle form of death had rained down on an unsuspecting population on May 19, 1953, and other days during that decade, the mistakes and subsequent cover-ups assumed the proportions of a major crime committed by the federal government against its most trusting citizens.

Shot Harry was the ninth in a series of eleven tests of nuclear devices conducted at the Nevada Test Site that spring. Operation Upshot-Knothole was an extremely important test series. These were the Cold War years, and the United States was racing the Soviet Union toward development of a hydrogen bomb. The test schedule was crowded, and the test site management—and all the others who participated in the series—were weary by mid-May. The circumstances were ripe for a disaster.[2] From the start there were difficulties with the test of the 32-kiloton nuclear device perched atop a 300-foot aluminum tower on Yucca Flat. The test had originally been scheduled for May 2 but had been postponed when Shot Simon contaminated the firing area on April 25, thus delaying the installation of some sixty experiments for Harry.[3]

On May 14 thirty-seven members of Congress, including freshman congressman Tip O'Neill from Massachusetts, arrived at the

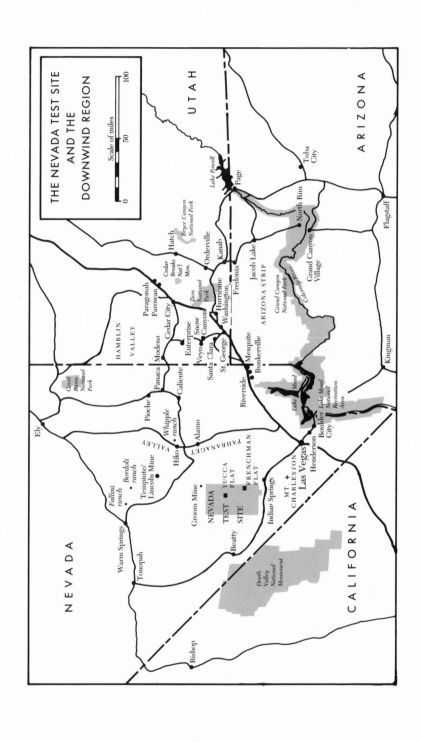

THE NEVADA TEST SITE
AND THE
DOWNWIND REGION

Scale of miles

100

50

0

UTAH

ARIZONA

NEVADA

CALIFORNIA

Lake Powell

Page

Tuba City

Flagstaff

Kingman

North Rim

Grand Canyon Village

Grand Canyon National Park

Colorado River

ARIZONA STRIP

Jacob Lake

Fredonia

Kanab

Orderville

Hatch

Bryce Canyon National Park

Cedar Breaks Nat'l Mon.

Paragonah

Parowan

Cedar City

Zion National Park

Hurricane

Washington

St. George

Santa Clara

Snow Canyon

Veyo

Enterprise

Modena

Mesquite

Bunkerville

Riverside

Lake Mead

Lake Mead National Recreation Area

Boulder City

Henderson

Las Vegas

MT. CHARLESTON

Indian Springs

FRENCHMAN FLAT

YUCCA FLAT

NEVADA TEST SITE

Groom Mine

Beatty

Tonopah

Warm Springs

Bishop

Ely

Great Basin National Park

HAMBLIN VALLEY

Panaca

Caliente

Pioche

Whipple ranch

Alamo

Hiko

Tempiute/ Lincoln Mine

Bordoli ranch

Fallini ranch

PAHRANAGAT VALLEY

Death Valley National Monument

test site and began to wait. Also on hand were nine hundred servicemen who were to witness the shot from trenches 4,000 yards from ground zero, twenty-six sheep that were closer to the tower, and an unknown number of civilian and military support personnel. The test was again postponed on May 16 and delayed yet again in two 24-hour increments because of unfavorable winds. The congressional contingent dwindled to twenty-three as impatience mounted.[4]

Weather conditions were far from ideal for Harry. A low-pressure trough was positioned off the West Coast and was pushing a series of fronts to the east. Weather balloons were released throughout the night to determine the direction of the winds aloft. The wind runs showed that the varying direction and speed of the winds at different altitudes would disperse the fallout cloud in an arc extending from north of Cedar City, Utah, to Lake Mead in Nevada. Just before the shot, the winds began to veer so that more of the fallout would be concentrated on St. George, Utah.[5]

The test manager's advisory panel, which consisted of various specialists and was headed by a test director, who was a scientist, said proceed. The test manager, an AEC employee who had the final responsibility for the decision, agreed. Harry was detonated at 5:05 on an overcast morning.[6]

Milliseconds later Harry, with a force nearly three times greater than the atomic bomb dropped on Hiroshima, burst from the instantly scarred floor of Yucca Flat. The roiling, blinding white-orange-red stem of the fireball had a nascent cap at its top and a burgeoning gray dust surge at its base. Harry, which later became known as "Dirty Harry," was distinguished from other shots by its rawness and unevenness—the qualities of an adolescent.[7]

The loose soil and rock of the desert floor, the tower, the cab that contained the device, and its casing were vaporized. As the fireball rose and began to cool, the gaseous vapors condensed. A strong updraft, termed an *afterwind*, sucked up the remaining loose debris from the desert floor and mixed it with the vaporized material and with the radioactive products that had escaped fission. In such a manner was fallout created. The particles varied in diameter

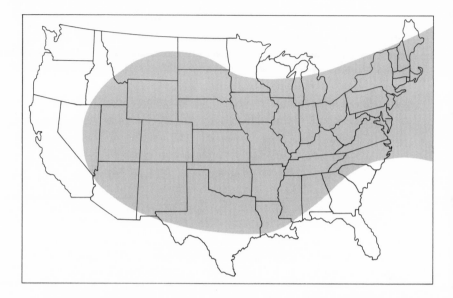

General Path of Fallout Across the United States

from one millionth of a meter to four-tenths of an inch. The heavier particles descended first.[8]

Ten minutes after the detonation, the rising cloud collided with the troposphere and spread out at the 42,000-foot level while the bottom of the cloud held steady at 27,000 feet. The cloud and stem retained their characteristic mushroom shape for a short period and then began to disperse. As the cloud drifted east, it retained a tinge of pink: the remains of nitrous acid and oxides of nitrogen formed in the intense heat and radiation of the fireball.[9] It was a scene of terrible and quite deadly beauty.

When Harry wandered offsite, it began to cause problems. The radio began to crackle with messages from the test site management to the offsite radiation monitors, collectively code-named "Hickory Stations," and individually referred to as G-50, A-68, or a similar letter-number combination so as to confuse people who might be monitoring the broadcasts.[10]

Fifty minutes before Harry was detonated, William S. Johnson, who was in charge of offsite radiation safety, had advised the U.S. Public Health Service (PHS) monitors that the "area of interest" was over a relatively unpopulated area southeast of ground zero.[11] An hour and seventeen minutes after the detonation, Johnson began activating monitors in other areas.

At 6:22 the message went out to the monitor at the Groom Mine, adjacent to the northeast edge of the test site, to "have residents stay indoors until further notice."[12] The mine owner, Dan Sheahan, noted in his diary: "Another very large tower shot atomic test set off at about 5:05 A.M. today. It was the last tower left standing (aluminum tower). In size, it appeared about as large as the April 25th shot. The high and middle part of the cloud went eastward toward Mesquite. The lower tail circled us clockwise from west to north. Fallout here at 6:15 to 6:19 A.M."[13] Like other offsite inhabitants that day, the Sheahans and the half dozen or so other residents of the mine were notified to take cover too late.

The miners and their families lived closest to the ground zeros on Frenchman and Yucca flats. From the old lead and silver mine, which consisted of about a dozen cabins and outbuildings on the flank of Bald Mountain, there was a direct view toward Yucca Flat, thirty miles in the distance. The test series that spring frightened the miners, who were veterans of two previous years of testing. The explosions were the biggest yet. An earlier shot had broken thirty windows, blown open doors, and stripped the sheet metal from the sides of buildings. Sheahan was worried. He later remembered, "Those tests that I am telling you about in 1953 were something to see. They were terrifying."

The miners were occasionally told by AEC personnel to evacuate the area because of the possibility of fallout, but they were assured that there was no danger. Then why should they have to leave? This paradox, which was never answered satisfactorily by the AEC, puzzled others throughout the years of atmospheric testing.

Sheahan noticed that the hides of deer, horses, and cattle that grazed in the area were speckled with burn marks, and there seemed to be fewer rabbits where the fallout was heaviest. A group of researchers from the University of California at Los Angeles, who were under contract to the AEC, fled the area when they thought they had accumulated dangerous doses of radiation. An army officer warned a private not to drink the water at the mine.[14]

Sheahan vividly recalled one shot in particular:

> There was a small thunderhead right over the top of the tower at about the 10,000-foot level. As the bomb went off the mushroom cloud picked up this thunderhead and made a beautiful thing out of it, although it was a dirty thing below. This cloud then soared right over our property and joined some other clouds and it started to rain. In the rain there were some large particles of fallout material, many of which were iron. We picked them up with a magnet. And since we knew that iron [steel] would be part of the tower, and we had been advised that the tower would probably contain a very dangerous material, probably even containing alpha as well as other rays, we were very worried about it.[15]

Shortly after the 1953 test series, Sheahan notified the AEC that he wanted to sell the mining property that he had worked since 1919. He was told by AEC officials to submit a claim and it would be considered. The claim was turned down. Sheahan then sued the AEC for property damage, and his wife sued the AEC for the cancer she had contracted. Neither was successful.[16]

Meanwhile, another portion of the fallout cloud that was headed southeast toward Mesquite, Nevada, passed over the southern end of the Pahranagat Valley. A brisk southerly wind then arose and pushed the cloud up the lush valley, a desert oasis watered by three springs and dotted with numerous ranches.

At the same time that those at Groom Mine were being told to take cover, Johnson ordered monitor C-84 "to contact Mr. Sharp and set up a precautionary roadblock at Alamo. Gently warn motorists proceeding south and monitor cars from the south. C-81 is coming to assist." Ainslee Sharp was the local deputy sheriff, and Alamo was a small community midway through the valley.

Radiation readings were taken on cars traveling north from the main St. George–Las Vegas highway that had passed through the fallout. Those cars that showed high readings were washed. It was not until 7:40—again, too late—that cars were stopped from proceeding north to Alamo.[17]

The deputy, whose horse was named Fallout because of the radiation burns on its back, was mad. Sharp did not like doing the AEC's bidding, nor did he trust the assurances of the AEC monitor that everything was all right. The monitor's radiation detection device, Sharp could see, was going offscale. The monitor later reported to his superiors: "Several of the local people at Alamo were greatly concerned over the tests. They expressed an opinion that they were being used as guinea pigs. The local people tend to disbelieve the AEC press releases."[18]

By this time in the morning, AEC public relations officials had cranked out four press releases concerning Shot Harry. The first said that the shot had been detonated. The second noted that Harry had been heard distinctly in Las Vegas and in Bishop, California, and had been recorded on seismographic instruments in Pasadena. The third was a lengthy statement by Representative Hugh D. Scott of Pennsylvania, who was the spokesman for the congressional observers. Scott said they had been advised that the shot was "very successful."[19]

The announcements then became less euphoric. The fourth press release, issued at 7 A.M., noted that the cloud had moved offsite and that there might be fallout. Motorists were warned to close their windows and air vents. Checkpoints would be established along major highways. The press release ended, "If there is fallout, it will not exceed the non-hazardous levels experienced after the April 25 shot."[20] This statement was both false and mistakenly clairvoyant. The radiation levels from Shot Simon had been hazardous, and they were exceeded by Harry. There was no way that even the all-knowing AEC could foretell what would happen.

Louise Whipple was driving the school bus to her teaching job at the Alamo school when Deputy Sharp stopped her and said, "You'll have to wait. We have orders. We have to wash all those cars coming from the south." With her on the bus was her youngest son, Kent.

A dark purple cloud hung over the small community, the closest to the test site. The air smelled putrid to Mrs. Whipple. She asked why they were being detained right in the middle of the fallout; it did not make sense.

Sharp stomped off.

But Mrs. Whipple did not choose to futilely protest the situation, as Sharp did. She said, "If the government had to have this for the defense of our country, I accepted it." Despite the dark clouds, the widow said, there were constant reassurances from the government "that things were okay and there was nothing to worry about."[21]

Mrs. Whipple believed the government then, but as the tests continued through the years, trust turned to doubt and anger. There was the time that Keith, her oldest son, and she noticed the AEC bus parked near their ranch. Keith asked the driver why he was parked there and was told the bus would evacuate the ranchers in case fallout came their way. Keith laughed. Most of the ranchers were scattered all over the county minding their livestock on horseback. When the fallout did come their way, the bus left without any passengers. The ranchers surmised that the driver and his crew had panicked.[22]

In 1958, as part of the continuing AEC public relations program, Mrs. Whipple and other ranchers who lived adjacent to the test site were invited to tour the facility. The fourteen men and women from the Hiko, Alamo, Ely, Tonopah, and Lincoln Mine areas were treated royally. After touring the site they were given a two-hour briefing. The emphasis was on offsite safety. One angry rancher, Paul Stewart, asked why he had suffered livestock losses during the 1953 tests. Mrs. Whipple said that Howard L. Andrews, a biophysicist from the National Institutes of Health who served on the test manager's advisory panel, had stated, "We admit we didn't know what we were doing then, but we do now." Mrs. Whipple never forgot that statement and recalled it twenty-two years later in a sworn deposition. The bland AEC press release that purported to cover the visit did not mention the incident. Stewart later died of cancer.[23]

On the day of Shot Harry, Keith Whipple was mowing hay when a strong wind came up from the south. The radioactive dust arrived with the wind, and Whipple could feel his exposed skin being burned. It stung and raised welts like so many mosquito bites. Whipple left off mowing and went to help a friend at his service station. The two men washed the vehicles that had passed through the fallout and were burned by the dust particles that had collected on the cars.[24]

Others that day in the downwind region had similar complaints. Headaches, fever, thirst, dizziness, loss of appetite, general malaise, nausea, diarrhea, vomiting, hair loss, discoloration of fingernails, hemorrhaging, and burns are all symptoms of radiation sickness and indicate exposure to relatively high doses of radiation. Nausea and vomiting can occur at levels as low as 50 rads, but more commonly take place in the 100 to 150 rad range. It takes more than 100 rads, possibly between 300 and 500, to produce red skin and hair loss.[25]

Later-developing cancers—the first could show up after one year, the last at an undetermined time in the future—can be induced by exposure to as little as a few rads.[26] Generally, the greater the dose, the greater the chance of contracting cancer. But exposure to radioactive fallout does not automatically produce cancer, which can skip one person and land on another with no rhyme, little reason, and no certainty about its cause.[27] Radiation is invisible and leaves no internal fingerprints.

There are other causes of cancer, including stress, exposure to industrial chemicals, and genetic factors. One in three persons in this country contracts the disease; one in four dies of it. The statistics for the downwind population, which was predominantly Mormon and rural, are somewhat less than the national average, since the Church of Jesus Christ of Latter-day Saints advocates abstention from the use of tobacco, alcohol, tea, and coffee—all possible causes of cancer.[28] The apparent crime would be difficult to prove.

The AEC health guideline for the offsite population at the time was 3.9 roentgens, a roentgen being roughly the equivalent of a rad. Both are units of measurement for radiation. This amount of radiation could be accumulated over thirteen weeks, a period that conveniently covered a test series. It was a general guideline that could be, and was, exceeded if AEC personnel decided it was necessary to conduct more tests.[29] There was little empirical data to support the 3.9 figure, other than some experiments with mice. The number seems to have been pulled out of a hat in 1952 by Gordon Dunning, an AEC scientist. It was a matter of balancing benefits against risks, said Dunning.[30]

But the Whipples and the other innocent inhabitants of the region had no knowledge of these figures as the bruised mass of the cloud inexorably moved eastward.

The fallout descended not only on humans that day. It also fell on thousands of sheep that were being trailed back to Cedar City from their winter range in the Nevada desert. The animals had also been exposed to the fallout from an earlier shot. The muzzles and backs of the grazing sheep were burned, and newborn lambs and ewes died in large numbers. The people who tended the animals were also exposed to the fallout.[31]

In the Hamblin Valley, just inside the Utah border and west of Cedar City, Elma Mackelprang was caring for some ewes and newborn lambs. Because it was a chilly morning, her three children remained in the warm cab of the pickup while she gave the sheep some water. As Mrs. Mackelprang worked, a fine ash, similar to the ash produced by a forest fire, settled about her. That afternoon Mrs. Mackelprang came down with a fever, nausea, and diarrhea. Her exposed skin was burned, and it peeled countless times. Three weeks later her hair began to fall out, and she continued to lose hair by the brushfuls until she was almost completely bald. She also lost her fingernails and toenails, which, along with her hair, eventually grew back. Her children, who had remained in the pickup, were not visibly affected by the fallout. Mrs. Mackelprang, who was twenty-nine years old, became quite nervous. The AEC subsequently told a curious local official that the woman had undergone a hysterectomy and that that explained the symptoms.[32]

The ominous cloud approached St. George, the largest community in the downwind region, with some 5,000 inhabitants. Frank Butrico, a young Public Health Service officer just out from New York City, was the offsite radiation monitor on duty in St. George. When he had arrived in town two days earlier, Butrico had checked into the Big Hand Motel and had notified Mayor Joseph T. Atkin of his presence.[33] He then took his bearings. One structure dominated St. George. The gleaming white Mormon temple stood above the leafy green trees and lawns of the town and challenged the surrounding red cliffs of the nearby desert for supremacy.

St. George was a model of small-town orderliness and, not coincidentally, the center of Mormonism for southern Utah and portions of the two adjacent states, where more than 90 percent of the inhabitants practiced that fundamentalist religion and its accompanying lifestyle of innocent belief. Mayor Atkin foresaw rapid growth for St. George in the early fifties and was determined that the growth would be orderly. Building codes and zoning ordinances were imposed, though there were mixed reactions to the new regulations. A history of the town noted, "It was not an easy task to remove all the animals from the city, but it was done."[34]

The residents of balmy southwestern Utah lived a great deal of their lives outdoors and were unusually self-sufficient. They were mostly dependent on the food that they raised themselves. A survey of their lifestyles during the years of atmospheric testing (1951 to 1962) showed that 55 percent obtained milk from their own cow, 44 percent drank milk with every meal, there were children in 45 percent of the homes, 23 percent breast-fed their infants and 22 percent fed their children fresh cow's milk, 56 percent obtained their drinking water from a spring, and 65 percent grew leafy vegetables.[35] Their self-sufficiency served as the seed of destruction for some. A few years after the height of the fallout in 1953 it was determined that eating radioactively tainted food products, as well as direct exposure, could induce cancer.

The subject of radioactive fallout was absent from the local literature during the fifties, which is not surprising for a region that saw itself as being on the way up. *U.S. News & World Report* dubbed St. George "Fallout City" a few years after the 1953 tests, but locals referred to the region as "Dixie," the "Land of Color," or "Color Country." Arthur F. Bruhn, a science instructor at Dixie Junior College in St. George, wrote a photographic guide to the region. Such a guide was necessary, he said, because of "the inadequacy of words to portray the beauties of Southern Utah's Land of Color."[36]

For outsiders, it was a place to pass through quickly on the way to somewhere else. St. George stood astride the main Los Angeles–Las Vegas–Salt Lake City highway and was the gateway to numerous national parks and monuments on the Colorado Plateau to the east. To the west was the other world. The descent from St. George was into the Nevada desert, Las Vegas, and eventually southern California. To prepare for this ordeal on wheels and to welcome the returning traveler, St. George Boulevard was lined with numerous auto courts, cafes, and gas stations. Because of the increased traffic through town on Highway 91, the main street was widened in 1953.

There were other events that spring. The American Cancer Society held one of its regular cancer clinics there in February, and the 167 attendees were urged to get regular checkups. Mayor Atkin proclaimed April cancer control month. In early May a civil defense coordinator was named for the town. The idea was to plan for the expected influx of refugees when atomic bombs dropped on coastal cities. What happened instead was a direct attack on the desert oasis, and for this there was no preparation.[37]

The superintendent at nearby Zion National Park noted that May was one of the windiest months in recent memory.[38] Norma R. Lyman, a columnist for the *Washington County News* in St. George also took note of the "very changeable weather in Dixie this spring" and wrote shortly before Shot Harry: "Atomic detonations are becoming so commonplace with us now that few of us in the area even raise an eyebrow when someone says, 'That must have been a big one. It rattled my dishes and opened my door.'" When vehicles were checked near St. George for radioactivity following the April 25 shot, the weekly newspaper stated that the AEC had "emphasized that persons, animals or crops exposed to the material were in no danger."[39]

During the week of Shot Harry, seven marriage licenses were taken out in St. George, thirteen births were announced, police warned about dogs that had not been vaccinated for rabies, high school seniors were preparing for graduation, and the local National Guard unit was readying for summer camp. The lead editorial in the newspaper noted that the soldiers being killed in Korea were much younger than those who had been killed in World War II. A plea for rain came from the correspondent out on the Arizona Strip, a desolate stretch of land between the Utah border and the Grand Canyon. If the scattered ranch families in the outback did manage to get to town, they could see Humphrey Bogart and June Allyson in *Battle Circus* at the Dixie Theater.[40]

On the morning of May 19, Frank Butrico monitored the AEC radio frequency and heard a great deal of chatter about roadblocks being established to the southwest toward Las Vegas. The residents of Mesquite and Bunkerville were then warned to go indoors.[41]

The 7:12 A.M. entry in the communications log at the test site noted: "St. George station operator reported arrival of roadblock team and instructed to set up precautionary roadblock at St. George on Highway 91 and report by 0800." Butrico reported that the roadblock, whose purpose was to halt traffic through the Mesquite-Bunkerville area, was in place at 7:45 A.M. The team consisted of Butrico and a policeman.[42]

The monitor was not told that the main fallout cloud was headed directly toward St. George, for the simple reason that the AEC did not know where and when the fallout would occur until a high reading was obtained by someone like Butrico.[43] The warning system was akin to shutting the barn door after the animals had escaped.

After helping man the roadblock, Butrico got into his car and drove to the center of town, where he took radiation readings. Shortly after 9:15 A.M. his instrument went offscale, indicating a reading of 7 roentgens. Butrico rechecked the accuracy of his instrument. Same reading. He telephoned the test site and talked to Johnson, who was surprised by the news. Johnson asked Butrico to recheck the instrument and take another reading. Same result. Call back in fifteen minutes. Butrico called. He was asked to take yet another reading. For the third time the monitor went through the procedure of rechecking the instrument and taking a reading.[44]

The bad news would not go away, so at 9:30 Johnson relayed the order of the test director, Alvin C. Graves, to advise residents to go indoors.[45] But how to warn them? The AEC had no plan for such an emergency. Neither Johnson nor Butrico knew whether there was a radio station in St. George, and if there was, where it was located. Precious minutes were passing. Fallout was dropping invisibly and silently to the ground. The monitor hurriedly went looking for the mayor and found him in his office.

Mayor Atkin knew what to do. He called the nearest radio station in Cedar City, fifty miles north. The gist of the message the mayor relayed to KSUB was that he had been told by AEC officials that fallout was occurring in St. George and that the authorities requested that residents of the town take cover and remain indoors until further notice.

The announcement was broadcast at 10:15 A.M.[46] No reason for the request was given, nor did it cover any of the other towns, hamlets, and isolated mines and ranches where fallout descended that day. Some heard the radio message and heeded it, others heard it and went about their business, and still others did not hear it at all. In any case, the announcement was too late. Butrico had obtained his high reading an hour earlier. Years later Johnson recalled in a workshop for the offsite monitors sponsored by the test site management: "I think we have to admit that we got the people in St. George indoors too late in this case. Just like we stopped the cars too late."[47]

Two state health officials who did not hear the warning were driving from Cedar City to St. George. E. Elbridge Morrill, Jr., the director of the Utah Division of Occupational Health, and Vic Pett, the chief safety officer, were about fifteen or twenty miles from St. George when the highway was suddenly enveloped by a thick dust cloud and the visibility cut to near zero. Fearing a collision, Pett stopped the car on the shoulder of the road and said, "Let's get out of here!"

The two men fled the car and crouched beside the main Las Vegas–Salt Lake City highway. Soon they heard the screech of brakes. Three cars collided, and the first in line lightly tapped their vehicle. The men climbed back to the car. Morrill, who was on his way to survey uranium mines in southern Utah, had left a radiation measuring device on the front seat. He tested it, and the needle went offscale on all three of the instrument's ranges, indicating a reading of 100 roentgens.[48]

JoAnn Taylor was approaching St. George from the west. She, too, did not hear the announcement. The young woman had just driven a friend to nearby Santa Clara and was returning when she passed through a heavy rain shower. Butrico also thought it may have rained that day in St. George. Rain brought fallout to earth, a condition known as a *rainout*. The AEC maintained that it had not rained within 1,000 miles "downstream" of the test site. Its successor agency, the Department of Energy, later said that it had not rained that day south of Cedar City.[49]

A few weeks earlier, Miss Taylor, who was a student at Dixie Junior College in St. George, had gone on a geology field trip led by Arthur Bruhn. One of the purposes of the outing was to observe an atomic explosion from a viewpoint in the Beaver Dam Mountains, about a half-hour drive west of St. George. Bruhn frequently watched the shots and sometimes required his students to do the same. The AEC assured Bruhn that no danger was involved in such an exercise. A fallout cloud came over the mountains, and Miss Taylor was badly burned and subsequently lost most of her long hair. When she combed it, the skin came off her scalp in strips. On the day of Shot Harry, JoAnn Taylor was halted at the AEC roadblock at the west end of town after passing through the rain shower. A monitor took a reading of her car and directed that it be washed at a nearby gas station.[50]

Kent Anderson lived a block from the highway. He became curious when he saw cars, trucks, and busses clustered at the service station. Thinking there had been a big accident, he walked down to find out what had happened. Anderson was told that the vehicles were being washed because they had passed through the atomic fallout. It was the first he had heard of it. His wife, who was outside gardening that day, was burned across her back where her blouse and jeans parted when she bent over.[51]

From what Frank Butrico could observe, the radio announcement was only partially effective in St. George. The monitor and a police officer cruised up and down a few streets and warned those persons they found outdoors. They passed a schoolyard and saw children playing outside. Butrico later stated that it was "quite obvious that the school hadn't gotten the message because the children were out there in recess."

A few citizens asked Butrico what to do, and a man in a pickup truck found the monitor and told him that his goats had turned blue. Butrico confirmed the discoloration, and other AEC officials later told the farmer that the blue color was caused by the goats' rubbing against the zinc coating of the fence. There was no mention of fallout being a possible irritant.

The monitor took cover for a short time and made some visual observations. He watched the fallout cloud, which was very dark, move slowly over the town and on toward the east, where there were no offsite monitors in the small communities.[52]

One of those communities was Kanab, Utah, where the cancer prevention clinic had also been held earlier that month. The big news in Kanab was the possibility of oil and gas leasing on the nearby Arizona Strip. Government men were in the area to do some surveying, perhaps in conjunction with the impending energy developments. Savings bond sales were up, and 123 pints of blood had been donated for the troops in Korea, an amount that compared very favorably with larger cities in Utah. A movie called *The Darkening Shadow* was shown that week at the Kanab Ward Chapel. Seven great-grandchildren of Richard Smith Robinson, a Mormon patriarch at the turn of the century, were among the twenty-six students to graduate from Kanab High School.[53]

On May 19, Elmer Jackson left his home in Kanab and drove to one of his grazing allotments on federal land in the Arizona Strip east of St. George. Jackson, a former mayor of Kanab, was going to move his livestock from one stock pond to another. While working his cattle on horseback, the rancher looked up and saw a dark cloud hovering over the hills just to the north. The wind changed and swept the cloud down on him. A light ash settled around a curious Jackson, who dismounted to inspect what looked like dirty snow. It burned. Jackson thought he might be in trouble, so he let the cattle run loose and got back on his horse. Somehow his hands had wiped his eyes. The rancher's eyes burned, and he was partially blinded. Fortunately, his horse knew the way back to the truck, and

Jackson managed to drive home. The reddish-brown splotches on his exposed skin, which resembled old leather or burnt flesh, never healed and peeled repeatedly. In 1968 Elmer Jackson was diagnosed as having cancer of the thyroid, which can be caused by radiation. Before he died in 1972, the rancher filed a claim for damages with the AEC. It was denied.[54]

At 11:25 A.M. the decision was made to give the all clear for St. George after Butrico called Johnson at the test site and reported: "The natives are getting restless." The mayor relayed the all-clear message to the Cedar City radio station.[55] The fifth press release was issued by the AEC public information office in Las Vegas at 11:30. It took note of the situation in St. George and stated: "None of the fallout levels reported, either on the highway or in St. George, are hazardous."[56]

A few minutes later Butrico met some worried AEC and military personnel who had flown from the test site to St. George to visit a uranium mine near Zion National Park where miners were complaining of headaches, nausea, rashes, and vomiting. After a brief talk with Butrico, who outlined the situation in St. George, the officials departed. The miners told the officials that they had heard the radio warning for St. George but had not taken cover. Their radiation detection devices went off all three scales. After becoming sick, the miners and their families left the mining camp. The next day the officials told the miners that they were suffering from "some gastrointestinal disturbance and the fact that it happened that morning could be coincidental." Anxiety over the radio announcement was another possible cause of the symptoms, they said. Since the local doctors whom the miners and their families consulted could not enlighten them, the miners accepted the officials' explanations.[57]

Butrico, who did not wear a hat, became concerned for his own safety when he got high readings from his hair. Because of his training and a briefing at the test site, he knew exactly what to do. When he had some free time around noon, he went to his motel room. He showered and repeatedly shampooed his hair. Butrico had not brought extra clothes with him because he expected only a short stay in St. George, so he purchased new clothes and discarded the old ones. Although outside for most of the morning and in the midst of measurable fallout, Butrico suffered no symptoms of radiation sickness after taking these precautions.

The monitor was not told by his superiors, nor did he take it upon himself, to notify St. George residents of these simple decontamination procedures. What he was instructed to say was, "We got some readings. . . . We know what happened. . . . We think it's not of much concern."[58] A few days later at the test site Butrico was to discover why the population was not instructed in what to do.

All the other monitors were recalled to the test site that afternoon, but Butrico was ordered to stay in St. George to answer questions and calm the populace. Remaining in the area was not an assignment that he relished, because of the high radiation readings. Butrico regarded his job in St. George from that point on as being a "PR thing."[59]

The events of May 19, 1953, were to become the worst case of fallout descending on a concentrated population of civilians as a result of the weapons tests at the continental site. What did the Southwest, the nation, and the world know about Shot Harry at the time? They knew what the AEC chose to relate publicly, not what took place within the dense shroud of secrecy that obscured the true nature of nuclear health matters. Public relations legerdemain and "top secret" security classifications hid the worries that were to ooze out years later.

As the age of television was just beginning to dawn, most of the public learned about Shot Harry and its aftermath via the newspapers. The small weekly papers that served the immediate downwind region used the AEC assurances almost verbatim, as did the larger metropolitan dailies. The *Washington County News* noted, "Despite the precautionary measures, the AEC insisted that 'radiation had not reached a hazardous level.'" To the north in Cedar City, an editorial in the *Iron County Record* cited the same statement about the absence of hazards from the May 19 AEC press release and added that Dr. Gordon Dunning of the AEC had said that the radiation levels were within the safety guidelines. The *Kane County*

Standard in Kanab ran a letter from state health officials, who were dependent on the AEC for information, stating that three children who reportedly had been suffering from radiation sickness were instead the victims of measles.[60]

In the regional centers of population the *Las Vegas Review-Journal* cited the statements of the governor, the local congressman, and the manager of the chamber of commerce, all of whom favored continued testing. The newspaper then editorialized: "We like the AEC. We welcome them to Nevada for their tests because we, as patriotic Americans, believe we are contributing something, in our small way, to the protection of the land we love. None should be able to deny us that privilege, so long as we do not encroach on the rights of others." The Mormon-owned *Deseret News* in Salt Lake City noted, "Atomic Energy Commission information men yesterday were working like proverbial one-armed paper hangers trying to deflate the mass hysteria which reached all the way back to the nation's capital regarding radiation fallout on the last atomic test." The rival newspaper, the *Salt Lake Tribune*, requested that future tests be conducted elsewhere, "preferably on some uninhabited island far out to sea."[61]

The out-of-state newspapers added little to the story, which was easily contained by the AEC. The *Los Angeles Times* noted that windows shook in Los Angeles from the blast. The *San Francisco Examiner*, using an International News Service story, cited the quote about no hazards. The *New York Times*, that journalistic model of thoroughness, initially used a United Press wire service story to report the test. It contained the same assurances and the incorrect statement that 5,000 persons had been forced to take shelter for three hours. The *Times* then sent its Los Angeles correspondent to Nevada to do a follow-up story, which was headlined: "Atom Test Studies Show Area Is Safe; Radiation Is Found Well Below Hazardous Level—Damage to Property Remains Small."[62]

The story, written by Gladwin Hill, was based entirely on information supplied by the AEC. Years later Hill, a veteran reporter who had covered numerous tests, recalled:

Except for what you could gleam from personal observations, informationally the AEC was the <u>sole source</u>, both *vis a vis* the reporters there and *vis a vis* the American people as a whole. The only things, practically, that anyone knew about nuclear detonations or radiation were put out by the AEC, and they were generally accepted as gospel.[63]

The *New York Times* story referred to the criticism leveled at the AEC by Utah congressman Douglas R. Stringfellow, who said his constituents were alarmed and promised a full investigation. The AEC promptly invited the congressman to witness the next test, on May 25, after which Brigadier General Kenneth E. Fields, a high AEC official, reported to AEC chairman Gordon Dean: "He got what he wanted and I think he will be very helpful to us." Stringfellow then publicly backpedaled and said that the AEC was doing everything possible to protect the offsite residents. He urged that the AEC's public relations program be strengthened and quoted the Bible thusly: "Know the truth and the truth shall make ye free."[64]

Utah senator Arthur V. Watkins wrote to Dean and said that he and his patriotic constituents were requesting that "extreme precautions and the greatest possible degree of care should be exercised in regard to future tests." Dean answered, "It is our sincere belief that the precautions of our Test Organization at every detonation are such that the possibility of any real danger to human health or livestock outside our Proving Ground does not exist."[65]

At the staff level, Gordon Dunning assured a St. George resident that "you and your family have not been exposed to dangerous levels of radiation." He outlined the radiation standard that he had established and downplayed the amount of exposure from Shot Harry.[66]

The public and private dialogues went on at different levels, and their content bore little relationship to each other, though the subject matter was the same.

The commission chairman and staff scientist did not publicly mention that, after hearing from Test Director Alvin Graves two days after Shot Harry that "mistakes in judgment are always possible, especially after a very tiring test series," the AEC had warned the test site management to be more careful because it feared that public opinion would force the abandonment of the continental test site. Everything should be done, said Dean, to avoid another fallout incident over St. George. The commission was told that the radiation measurements recorded in St. George were as high as 6 roentgens. But fallout was not the only concern. The AEC had to counter a large number of complaints from across the country that the Nevada tests were causing abnormal weather patterns that spring. As with the danger of fallout, the commission's position was one of denial.

Public health was of little concern compared to the perceived need for national security. "In the present state of mind of the public," said Commissioner Eugene M. Zuckert in the privacy of the commission's meeting room, "it would take only a single illogical and unforeseeable incident to preclude holding any future tests in the United States." The "psychological problem" and "public relations aspects" of the tests were discussed by the commissioners and the staff. The military believed that the commission was overemphasizing the dangers of fallout and had taken "extreme" precautions with the public, thus arousing its concerns. No, said Commissioner Henry D. Smyth, "The present program emphasizes the extreme care taken by the AEC in avoiding hazard to the public. This is believed to be the best way to create confidence and allay public fear."[67]

On May 21, at the suggestion of Dunning, who was at the test site, Butrico was told to obtain a sample of milk in St. George so that it could be analyzed for iodine 131, a radioactive isotope that was a by-product of fallout. He was not, however, instructed to obtain the sample in an ordinary manner. "Because of apprehension reported to exist among the residents of St. George, an indirect approach for the collection of milk and water samples was worked out," stated the communications log. Butrico did not obtain the milk directly from a dairy, as most residents who drank raw milk did, but instead purchased a bottle of pasteurized milk in a store. Such an evasive action was taken, said Johnson, so as not to "alarm an already worried community." Any radioiodine that may have been contained in the milk Butrico and the other monitors gathered was subsequently lost by a faulty laboratory processing technique.[68]

A postshot critique was held with high-ranking AEC officials back at the test site, and Butrico then found out why the residents had not been instructed in decontamination procedures. He was told that it was an unfortunate situation. As Butrico later paraphrased the response in sworn court testimony, the senior officials said:

> We have taken steps as best we could with reference to road-blocks and people going indoors. We have many inquiries about people reporting they are sick. But anything more that we try to do, the more we try to disseminate more and more information, let's cool it. Let's try to get this thing quieted down a little bit because if we don't, then it's likely that there might be some suggestions made for curtailing the test program. And this, in the interest of national defense, we cannot do.[69]

When discussing Shot Harry in June 1980 with Department of Energy officials who were trying to reconstruct the events on that fateful day, William Johnson said, "Yes. It was a rather shocking day to lots of folks." It was also shocking to Butrico to find out that Johnson had altered his report on Shot Harry to give a more favorable account of the day's activities. After reading the report for the first time at a similar session with DOE officials two months later, Butrico exclaimed, "If you're getting the impression I'm unhappy with this report, you're right. That's all I'd like to say for the record."[70] But the record of deceitful activities by the government during the years of atmospheric testing would emerge in time.

Thirty years after Shot Harry, the dry lake beds of Frenchman and Yucca flats were capped by low, gray clouds. Snow flurries, like fall-out, fell silently and unpredictably to the ground. The clouds were being driven eastward by the prevailing wind, as had those dark clouds produced by the mammoth atmospheric explosions of years ago. The rusted debris from the tests was strewn about the desert floor. There were mangled towers, a demolished bridge, electrical cables, the walls of a mock motel unit, and the wire cages that had once held animals for medical experiments. The desolate quality of the landscape was heightened by the pockmarked subsidence craters from the more recent underground tests, which gave the site a lunar aspect.[71]

Because of the secretive nature of the business that was conducted there, there was little to draw public attention to the site. No tours, such as those at nearby Hoover Dam, disturbed this preserve dedicated to perfecting the ultimate means of death. Nevertheless, the site, some seventy miles northwest of Las Vegas, was of great historical interest. The 1,350 square miles of the test site proper have been zoned into parcels where various means of mass destruction have been explored in a relentless manner since 1951. More nuclear weapons have been detonated here than at any other single location in the world. No other landscape in the Western Hemisphere has been so irradiated by man.[72] So it was not surprising that some of this deadliness leaked offsite.

Shot Harry contributed the most offsite contamination among the 105 announced atmospheric tests conducted at the Nevada Test Site between January 1951 and July 1962, when all testing went underground. In addition, 19 tests were held in unsealed shafts or craters. Most of the 124 tests resulted in radioactive fallout drifting offsite. Lesser amounts of radiation leaked from a few underground tests over the years.[73]

The population in the immediate area faced the most peril from this particular source of ionizing radiation, but fallout from Nevada, in decreasing amounts, also drifted across the country to the East Coast. No area in the country was immune to it.[74] Eventually the local fallout joined the global fallout—the product of the nuclear weapons detonations that was circling the world during the years of atmospheric testing.

It took time for the latency periods of the various types of cancers to elapse. Men, women, and children at the Groom Mine, Alamo, Hiko, Bunkerville, St. George, Cedar City, Kanab, Fredonia, and smaller communities and isolated ranches and mines died or were maimed by cancer in the years following the atmospheric tests. Undoubtedly, not all the cancers were caused by fallout, but just as surely, some were. The government, however, denied responsibility for all such injuries and deaths.

The only memorial to the victims of this distinctly American nuclear tragedy was a trial. Each plaintiff's confrontation with cancer was an intense story of extreme pain and suffering. Yet the sense that individuals were involved tended to become obscured by their cohesiveness as a group and the voluminous technical and historical testimony presented at the trial.

But it was people—indeed, a particular type of person—who suffered. The downwind people who lived in portions of Nevada, Utah, and Arizona were predominantly rural, Mormon, and Anglo-Saxon. They were unusually patriotic and innocent. They mostly endured rather than question or object. Besides, they were repeatedly assured by their government that there was no danger from the black clouds that passed overhead, and they believed what they were told by their government. They trusted. That was their downfall. Not until a quarter century after Shot Harry did they discover that they had been deceived.

The downwind civilians were also unique among weapons-related radiation victims. At Hiroshima and Nagasaki the two airbursts did not produce much fallout. What occurred near the Russian and Chinese test sites is not known with certainty. The Marshall Islanders were exposed to one intense dose of radiation in 1954. The servicemen who witnessed the Nevada tests, for the most part, were exposed only once and were instructed in safety precautions. On-site civilian workers may have been exposed more than once; but for them it was an occupational hazard. They had some knowledge of what the dangers were, film badges to measure the extent of exposure, decontamination facilities near at hand, and, ultimately, freedom of choice.

For the downwind population there was no choice and there were none of these precautions, except for a few film badges randomly handed out and haphazardly worn. But there were repeated exposures to radiation from atomic weapons and innocence—surely the scenario that would apply to most people in any nuclear holocaust.

To state unequivocally that one person or another died or was maimed by a cancer caused by radioactive fallout from the Nevada Test Site is impossible. Causing suffering and death are, to be sure, onerous crimes, but there is a broader and much more certain context in which to place this story. In the end, these people were betrayed by their government—the ultimate sin in a democracy. It was a crime of betrayal perpetuated in the name of national security, a concept that may have been relevant to begin with but that shaded into bureaucratic intransigence as the Cold War waned, the tests went underground, and détente emerged. The state of mind of the government not only froze in place, it fossilized as the years progressed and the world changed. At one end of the scale of injustice, this breach of faith could be viewed as an act of sustained stupidity, while at the other it resembled a perfidious act carried out by a government against its own people.

The folly is that the crime was committed against citizens of such overwhelming docility that, seen in retrospect, the deception was unnecessary. Had they been warned and instructed in the proper precautions, as those who worked in the weapons programs at the time had been, they would not have objected to continued testing. The truth would have worked.

But that was not to be, and ultimately the system failed. The problem was that the American system was not equipped to deal with the morality of the situation, nor was compassion a tool of government. The administrative, legislative, and judicial branches of government failed to find a solution. The dereliction of democracy in this case and what it portended in the era of Watergate, the Iran-contra affair, and so many other "affairs" was the ultimate fallout from this story.

What happened in this isolated region of spectacularly colored deserts, canyons, and mountains was Hiroshima and Nagasaki come home to roost in a much smaller but particularly haunting way. For the country that unleashed those two bombs on Japan and, along with the Soviet Union, retained the capacity to obliterate life on Earth, there was pause to think. And the conclusion might well be that there was no other such realistic example of the long-term effects of nuclear warfare that has been so well documented and that was so poignant and immediate as the fate of this one small portion of the Earth.

CHAPTER 1

THE DISCOVERY

THERE WAS an inexorable quality to the unlayering of the discovery. It began with an army veteran being diagnosed as having leukemia, led to the media, was picked up by a mother who was suspicious about the death of her son, was passed on to a widow who was familiar with the law, and ended in a momentous lawsuit. Along the way, the three branches of government attempted, in varying degrees, to deal with the avalanche of disclosures and the twin pressures of media attention and public concern. The attempt by the executive branch and its bureaucracy was minimal, Congress made more sustained efforts, and the judiciary could not avoid the issue. In the end, they all failed.

Paul Cooper was a veteran of Korea and two tours of duty in Vietnam, where he had been a member of the elite Special Forces, also known as the Green Berets. In 1957 Cooper, along with 3,200 other soldiers, witnessed Shot Smoky at the test site. Smoky was a mess. A radiation monitor later recalled, "Smoky, of course, was a good one. It dug in and did all sorts of things." After leaving the army as a highly decorated sergeant, Cooper was diagnosed as having leukemia in 1976; and the Idaho resident was sent to a Salt Lake City hospital. His claims for a service-related injury were repeatedly turned down by the Veterans Administration, so Cooper called KUTV, a Salt Lake television station. They ran a story on the hapless veteran. Then the *Deseret News* did the same.[1]

The newspaper, however, went one step further. It assigned its Washington correspondent, Gordon Eliot White, to do a follow-up story. White made the intuitive leap from soldiers to offsite civilians. With the help of statistics from the National Cancer Institute that showed a small increase in Utah cancers that seemed to match the period of atmospheric tests, he wrote a story to that effect in August 1977. The story was then picked up by other media outlets.[2]

In January 1978, Vonda McKinney's aged mother-in-law, Nora McKinney, of Globe, Arizona, sent the younger woman a letter and enclosed a newspaper clipping from the *Phoenix Gazette* of the previous December 16. Over the three-column story was the headline "Leukemia Link Confirmed; Nuclear Test Witnesses Sought by the Government." The Associated Press story out of Washington, D.C., related how Dr. Glyn G. Caldwell of the federal Center for Disease Control in Atlanta was attempting to locate veterans who had witnessed Shot Smoky and had participated in military maneuvers afterwards. Caldwell was studying whether there was an excess of leukemia among these former servicemen.[3]

The old woman wrote, "I am enclosing a clipping I got from the paper about leukemia. As I'm sure I told you, I had felt Lenn got it from exposure to radiation when hauling across Nevada when so many tests were being done. I believe it might be worth investigating from the Kaibab Lumber Company from their records, and it might be to your advantage to take it up. I didn't want to bother you about it when I sent the [Christmas] card so kept it to send later; and today is a letter day."[4]

At the time, Vonda McKinney lived in Holbrook, Arizona, where she was justice of the peace for Navajo County. Following the death in 1962 of her husband, Lenn, who had been a yard foreman for the lumber company and an independent truck driver, Mrs. McKinney had moved from Fredonia, where she had also been the justice of the peace and coroner in that hamlet just across the state line from Kanab. She knew lawyers and the law.

The day Mrs. McKinney received the letter, Alexander Russin, a friend and local attorney, walked into her Holbrook office. She showed him the note and the clipping and asked, "Would you be interested?"[5] Russin thought he might take the case and then sat on the matter until April, when he went to a conference of the Association of Trial Lawyers in Hawaii. The Holbrook attorney approached Dale Haralson, a successful Tucson lawyer who specialized in personal injury cases. Haralson had just given a talk at the conference. Russin complimented him on his presentation and asked if he was interested in handling three cases that had come his way.

He needed help, Russin said, because all three cases were complicated and would necessitate a considerable outlay of money before there was any possibility of a favorable judgment and compensation. The first case was an automobile accident, the second was a medical malpractice suit, and the third was Mrs. McKinney's suspicion that her husband's fatal leukemia had been caused by exposure to radioactive fallout from the Nevada Test Site.[6]

Haralson saw no problems with the first two cases. He was less interested in the third case, not having heard of such a situation before and suspecting statute-of-limitations problems—meaning that a lawsuit had to be filed within a certain time limit, which had seemingly expired in this case. But the Tucson lawyer had learned not to tell a fellow attorney that he would take his easiest cases, which were most likely to return a favorable fee, and not his more difficult ones.

It was by such careful assessments and hard work that Haralson had risen to the top ranks of personal injury lawyers in the Southwest within fifteen years. His life mirrored the life of the clients he came to represent in the radiation suit, a parallel that was almost carried to the ultimate limits. A deeply religious man who described himself as a Christian lawyer, Haralson grew spiritually with the third case. It was as if they were destined for each other.

Haralson was born in 1937 in west Texas and grew up on a 400-acre stock farm near Loraine. His mother was a schoolteacher, and his father was a farmer who sold automobiles during hard times. The whole family, including Haralson and his two brothers, helped with the farming. Outside the farm there was only school, the local Southern Baptist church, and an occasional Saturday night movie, until the local movie house closed. Like the Mormon villages to the northwest, there was a closeness about the community around Loraine. Life revolved around the family and the church.

In high school Haralson decided he wanted to become a lawyer. It was, he thought, consistent with what God wanted him to be. Such a profession would allow him to influence other lives and support church activities. Like the Mormons, the Southern Baptists tithed 10 percent of their earnings to the church. He graduated from the University of Arizona law school in 1963. Practicing law in government, a corporation, or a large law firm did not interest Haralson. He wanted to represent people who came from the same social and economic background as he did, and he thought he could communicate with a jury of such persons. This meant trial work, and particularly civil cases, which were more lucrative than criminal cases.

Haralson went into practice with an uncle in Tucson and took on cases that proved to be stepping-stones. He successfully challenged such giants as Sears Roebuck, General Electric, and the Southern Pacific Railroad in personal injury cases; and the rewards were plentiful, one case yielding a judgment of $3 million.[7] Haralson got a hefty chunk of that award.[8] These three cases demonstrated to him that it was possible to take an individual into the judicial system and prevail against much greater resources as long as he felt that right was on his side and competent witnesses and persuasive evidence were available. What he obtained from these cases, and others like them, was legal expertise, confidence, and money. In midcareer, Haralson was at the peak of his professional abilities and physical stamina.

After their return from the conference, Russin and Haralson talked further by telephone. Haralson remained doubtful about the McKinney case, but in order to take the responsibility off Russin's shoulders, he agreed to talk to the woman. Russin passed on the message.

Mrs. McKinney called Haralson in early May and briefly outlined her story. Haralson pointed out that there were problems with the case. Mrs. McKinney asked if she could come down to Tucson to talk about the matter in more detail. All right, said Haralson, without much enthusiasm. The woman had to drive through Phoenix to get to Tucson, and Haralson half hoped that she would stop in that larger city and talk to one of the many lawyers there. But Mrs. McKinney was not to be deterred. The two met in late May in Haralson's downtown Tucson office.

At the start of the meeting, Mrs. McKinney was skeptical of Haralson, and Haralson was skeptical of the case. She told him about the dark fallout clouds that had passed over Fredonia in the fifties and of the leukemia deaths that had followed in the small community—four within a short time where the disease had been unheard of before. That struck a responsive chord in Haralson. Fredonia, with about 600 inhabitants, was just about the same size as Loraine, where leukemia was also unknown.

By the end of the conference, Haralson was satisfied that there was the potential for exposure to radiation, although he knew little about the causes of cancer or the history of the nuclear testing program. He told Mrs. McKinney that even if he could prove that her husband's cancer had been caused by the fallout, the amount of the award might not be equal to the expense of conducting such a complex lawsuit. Then Mrs. McKinney mentioned the other women whose family members had died of leukemia in Fredonia during the 1960s. She asked if, from a cost standpoint, it would make sense to have them involved in the suit. Haralson said that it would make a difference, and he agreed to visit Fredonia if Mrs. McKinney could set up a meeting with the women. The attorney made it very clear, however, that he was making no commitment to take the case at that time.

Privately, Haralson's curiosity was aroused. He felt that something was there. The occurrence of so many cases of leukemia within a short period of time in such a place did not sound right. At the end of the meeting, Mrs. McKinney felt that she could trust Haralson and that he was a competent lawyer. A bond began to form between the two that was similar to the trust and respect that the other downwind clients later felt toward Haralson. Mrs. McKinney returned to Fredonia and spent a week talking to the other women. They agreed to meet with Haralson. The momentum had begun to gather.[9]

Mrs. McKinney and the women of Fredonia were the start of a phenomenon that held true on this issue for the whole downwind population. In Mormon society, women were supposed to be docile and passive. Yet it was the women of Fredonia and the other small towns and hamlets who, once they suspected a link between the tests and cancer, organized the challenge to the government and were the loudest and most vocal in their protests. At times these women sounded like the chorus in a Greek tragedy. At other times a lone female voice stood out. In comparison, the men of the region were mute. Perhaps it was the women who felt the keenest loss, the greatest abandonment by death. Perhaps it was the women who intuitively sensed the threat to future generations from the legacy of the tests. Perhaps the female role of passivity within this society was an illusion, or a media mistake. And perhaps it was the men, the ranchers who had to sell their products to outside markets and the businessmen who relied on the tourist trade, who felt their livelihoods most threatened by the disclosure that the area had once been poisoned.

In early June, Haralson flew north in a small plane that he piloted and picked up Russin and Mrs. McKinney in Holbrook. They proceeded to Fredonia, passing over the Grand Canyon on the way. The two lawyers met the women at the home of Jackie Lynch, who was the mother of Odessa Burch. Also on hand were Rose Mackelprang and Esther Adair, whose husband had died of a solid cancer. The Adair case was later separated from the remainder of the downwind cases because he had worked at the test site. The fourth woman who eventually filed suit, VeRene Tait, was at work. The attorneys visited her at home later that evening.

The women told their stories. A year before Lenn McKinney died, his daughter's best friend, Odessa Burch, had died of leukemia at the age of fourteen. Gayneld Mackelprang, the school superintendent, died in 1964, and Lavier Tait, who served with McKinney on the school board and at one time worked with him, succumbed to leukemia in 1965. The school, the church, the general closeness of the community, and the fallout all seemed to be common denominators. In the three families where a father died, fifteen children survived. Three widows and one bereaved mother formed the initial chorus.[10]

Haralson, who was measuring the effectiveness of potential witnesses, was struck by the credibility of these women. After listening to their stories, he outlined the difficulties of the case. The lawyer said he wanted to do more research before deciding whether to proceed. He promised to return soon to give them his decision.

And so began the scientific education of Dale Haralson. He began near the zero point and traversed the entire spectrum of health physics, nuclear physics, the medical aspects of cancer, and the history of the weapons testing program. Thoroughness is a Haralson characteristic.

The lawyer returned home and telephoned the University of Arizona, which is located in Tucson. He asked for someone on the faculty who was knowledgeable about radiation-caused cancers and was given the name of a radiologist in the medical school. Haralson made an appointment and went to visit the doctor, who told him he wanted no part in a lawsuit that might have an adverse effect on his profession, a reaction that Haralson was to encounter a number of times during his quest. But the radiologist did loan him a book, thus inadvertently providing the lawyer with two expert witnesses. The book, *Population Control Through Nuclear Pollution*, was by John W. Gofman and Arthur R. Tamplin.[11]

Haralson happened to mention what he was working on to his next-door neighbor, whose father, Heinz Haber, had written a children's book entitled *The Walt Disney Story of Our Friend the Atom*. The book was of little use to Haralson since it did not mention fallout.[12] However, Haralson was impressed by the Gofman and Tamplin book, and particularly the former's credentials. He called Gofman in San Francisco, identified himself, explained the purpose of his search, and asked if they could meet. The answer was yes, so Haralson flew to San Francisco and met with Gofman for three hours in a seafood restaurant.[13] "It was," wrote Haralson in a note for his files on the meeting, "the most exciting conference I've ever had in my law practice."[14]

John Gofman, like so many other scientists involved in the nuclear weapons program during the early years, was a product of the University of California at Berkeley, where he was involved in the discovery of uranium 233 and its fissionability while still a graduate student. During World War II, while with the Manhattan Project, which developed the first atomic bombs, Gofman found several ways to isolate plutonium. Following the war, he taught at the university and did research under Glenn T. Seaborg and Ernest O. Lawrence, both Nobel Prize winners and bulwarks of the early nuclear establishment. Gofman was both a nuclear physicist and a medical doctor, which was unusual for someone working in the weapons program.[15]

Through the 1950s and early 1960s, Gofman was very much a part of the nuclear club. In a 1957 speech that was widely circulated by the AEC to quiet its critics, Gofman said that the low levels of radiation being released by the tests were causing no harm. By 1962 he was an associate director of the Lawrence Livermore Laboratory, named after his mentor and one of the two weapons laboratories run by the University of California for the AEC—the other being Los Alamos. Gofman and Tamplin openly and forcefully challenged the AEC's position on low levels of radiation at the end of the decade. They became a cause célèbre within the agency's ranks. Both went on to careers in the antinuclear movement. In Gofman's case, there was a distinction to be made. He was an outspoken opponent of nuclear power plants, but he was also a strong supporter of the nuclear weapons program.[16]

The scientist gave the attorney a basic course in the dizzying complexities of radiation and health. He launched into an explanation of radioactivity, which is the spontaneous emission of alpha and beta particles and gamma rays from the disintegration of the nuclei of atoms. Alpha particles, he explained, cause internal harm. Beta particles are responsible for skin burns and hair loss and could be ingested through the food chain. Gamma rays, which are similar to X rays, can alter cells as they pass through the body.

Gofman then explained the difference between such radiation measurements as rads, roentgens, and rems. The roentgen, he said, is a unit of exposure to external gamma radiation and is roughly equivalent to a rad, which is a unit of absorbed dose of radiation. The rem is a unit of dose that takes into account the biological effect of absorbed radiation. Another way to put it is that a rad is a roentgen of radiation that has been absorbed by an organism and is roughly equivalent to a rem. There are whole-body doses and doses to specific organs.

The medical doctor then took over, with Gofman explaining how ionizing radiation causes cancer. At the time, Gofman was working on what would become a 900-page book, *Radiation and Human*

Health, which is a complete survey and explanation of what was known on the subject. On causation, Gofman wrote that not every person exposed to radiation will develop cancer. Instead, "a population exposed to a certain dose of radiation will show a greater incidence of cancer than that same population would have shown in the absence of the added radiation." The complex problem, then, is how to determine whether a specific person's cancer was caused by radiation.[17]

Gofman rattled off the names of a number of other possible witnesses who might be willing to challenge the contentions of the government. Among the most eminent, he said, was K. Z. Morgan, known as the father of health physics. Although "K. Z." had a tendency to ramble in a courtroom, Gofman said, "He's a very, very moral man and will be angered if he feels that there's an injustice being done to some people."

At the same time Gofman warned Haralson about one possible witness, Charles W. Mays, whose laboratory at the University of Utah depended on government funding. He should be used with care, said Gofman, since the information he learned from the plaintiffs might find its way back to the Department of Energy. Gofman warned, "We just have to speak cautiously when we're around him."[18]

The former government scientist suggested that Haralson search government document centers in Washington, Oak Ridge, and Las Vegas for information on the tests and their health effects. Haralson stopped at the Department of Energy's Las Vegas Operations Office on his way back to Tucson and made an informal Freedom of Information Act request. He was handed the offsite monitors' reports for the various test series at the Nevada site. It quickly became apparent to Haralson that the documentation was available to demonstrate that the hazards had been known to the AEC and that no meaningful warnings or protective measures had been taken for the safety of the offsite population.

The attorney now knew that the documentation and the expert witnesses were available to make a viable case. There seemed to be no legal impediments, such as the statute of limitations. Although there was a two-year limitation on filing suit under the Federal Tort Claims Act, the clock seemingly began running when there was knowledge of the cause, not when the event or the resulting deaths occurred. For the women he had talked to, this knowledge had come within the past year when there was extensive media coverage of the subject. Before then, they had believed the government assurances that there was no danger. With the additional clients, legal action made economic sense, and Haralson was now ready to proceed.

Another meeting was scheduled, and Haralson returned to Fredonia in June. After hearing Haralson's report, the women were not vindictive or bitter. They were just disillusioned with their government. Even after they understood the possible link between the tests and the deaths, suing the federal government was not an easy thing to accept. Mrs. Tait said, "There are many ways to look at the government. The government is the people. Yes, I have a reluctance to sue the people. They don't owe me a living."[19]

The contracts were signed, and the first claims seeking an administrative remedy—the necessary prelude before filing suit in court— were filed with the Department of Energy in Las Vegas in September 1978. Additional claims followed soon thereafter.[20]

The wide publicity that the initial filings received gave Haralson the first indication of the importance of the issues that the case had raised.[21] The dimensions of the case and his stake in its outcome were suddenly lifted out of the personal injury and wrongful death categories of a narrowly focused legal action for damages to become issues of national significance. The basic situation cried out for an answer to the question of what a government had done, knowingly or unknowingly, to its own intensely loyal citizens.

Another attorney became interested in representing downwind clients in the summer of 1978. Stewart L. Udall, whose immediate family's roots were in central and southern Arizona but whose four Mormon grandparents had lived in the downwind area, went to Washington in the mid-1950s as a congressman and remained there as secretary of the Department of the Interior in the Kennedy and Johnson administrations in the 1960s and as a lawyer-lobbyist in the

1970s. The fallout case and the possibility of other radiation-oriented litigation eventually brought Udall back to Arizona.[22]

In early July, Udall read a story by Bill Curry in the *Washington Post* headlined "The Clouds of Doubt Haunt the Mesas." Quoting people like Vonda McKinney, Curry asked: "A collective coincidence or individual and random tragedy? Or do their deaths, and those of so many others across this region of mesquite and mesas represent civilian casualties of atomic weapons—people killed by their government as it tested the arms that were supposed to protect them?"[23] Udall did not give the story another thought until a cousin, C. D. Stewart, called him and outlined the situation in Alamo.[24]

A meeting with the Washington lawyer was arranged in the St. George home of Mrs. Irma Thomas by Rose Mackelprang, who had been contacted by Udall. A confused Mrs. Mackelprang called Haralson. The Tucson attorney advised her just to listen and not to divulge that he was working on the case. Mrs. Mackelprang later sent Haralson a list of the participants. At the time, the two lawyers were competitors for the same block of clients.[25]

Udall flew to Las Vegas and, because he had forgotten about the time change between Nevada and Utah, he was an hour late getting to the meeting. There were eight people at the Thomas home. The next day he went to Alamo, where a similar meeting was held in the Stewart home. The lawyer listened and commiserated at both stops. He was sobered by what he heard and surprised that such people had come to distrust their government. Something had to be done, Udall felt, reacting with the instincts of a crusader.[26]

Not having practiced law in a courtroom for more than two decades, Udall approached the task differently from Haralson. Besides, they were different persons. Udall was impulsive, more a public personality, and less thorough. He had an interest in causes and saw himself as the champion of these downtrodden people. Being essentially a Washington animal, Udall's next action on his return to the capitol after the two-day visit was to seek out the media and help draft a petition for members of Congress from the western states to present to President Jimmy Carter. The petition, signed by the entire Utah delegation and the two Arizona senators, asked that the president investigate the matter.[27]

Udall returned to St. George in late September, a few days after Haralson filed the first claims and, accompanied by a drumbeat of publicity, held more meetings with the relatives of cancer victims.[28] A citizens group was set up to help coordinate the information now pouring in, and a local attorney, J. MacArthur Wright, joined forces with Udall and made the basement of his St. George office the nerve center of the operation.[29]

Udall was quoted often in local newspapers that fall. After interviewing 125 people during a four-day period in October 1978 in Wright's office, the Washington lawyer said the enormity of the situation was shocking, cancer rates were three or four times greater than normal, and the disease had cut through three generations of one family. An emotional Udall was shaken by these stories. "When these people tell you what this has done to their lives," he said, "it's a shattering experience. You can feel the depth of the tragedy that occurred here."[30] The media tended to focus on Udall and neglect Haralson, which fit their different styles.

It soon became obvious to the two lawyers, however, that they were duplicating each other's efforts and that they had some complementary talents. Haralson met with Udall in his Washington office in mid-October, and they agreed to join forces. According to the October 19 agreement, if the claimants were compensated, there would be an equal sharing of the fee. Wright, who had the primary responsibility for dealing with the clients, was to be paid from Udall's share, and Russin from Haralson's fee. (Russin had little more to do with the case and died shortly thereafter.) Generally, Haralson was to be responsible for developing the legal aspects of the case while Udall worked on a negotiated settlement and legislation in Washington. Decisions would be made by both "after appropriate consultation." In a position paper for use with the press, the two lawyers stated: "We are totally in agreement with respect to avoiding a confrontation and attempting to have these claims handled administratively."[31]

Over the next few years, the two lawyers had their differences but not to the point of seriously damaging the presentation of the case.

At one point, angered by the interference of Udall's Washington partner, Haralson fired off a letter to Udall: "Stewart, I have very strong feelings about how I try a lawsuit. I don't like more than two lawyers (preferably one, me) sitting at counsel table." [32] But the two were tied to each other, and Haralson would be joined at the counsel table by others.

The screening criterion for claimants was residency during the period of atmospheric tests in the downwind region where it could be demonstrated that the amount of fallout was relatively high. This was generally in an arc from north to southeast of the test site and out to a distance of some 250 miles. Medical records or a death certificate showing that the cancer was of a type that was thought to be caused by radiation was also needed. In December, Udall filed one hundred claims. [33] The filings would steadily increase.

Not everyone in the downwind region was pleased with what was happening. Local attorneys did not like the two outsiders trespassing in their bailiwick. Most of the criticism was directed at Udall, since he was the more visible and prominent of the two out-of-state lawyers. Udall, it was said, was too aggressive in the manner in which he attracted clients through news stories. What Udall was doing, several southwestern Utah attorneys implied, was running a well-orchestrated advertising campaign for his services, and this smacked of solicitation. The state bar association looked into the matter and dismissed the charges. [34]

One segment of the local population lashed out publicly at Udall. The merchants of the St. George area, which was receiving the brunt of the news coverage, were unhappy at the effect of the publicity on the tourist and retirement trade. The specters of radioactive fallout and cancer deaths were not good for business. A banker noted that home loans from southwestern Utah were beginning to be questioned, at least one major convention in St. George had been canceled, and tourists were staying away from the area.

Udall's motives were questioned in a letter from the executive director of the St. George Chamber of Commerce to local newspapers. Arthur B. Anderson suggested that the lawyer donate his fees to the affected families and that President Carter come to St.

George to make a public apology. Anderson was quoted in the local newspaper as stating at a chamber meeting that "his only concerns . . . were that tourism and other things have been affected by the publicity from the media and the attorneys involved." The chamber subsequently backed his position.[35]

At the end of 1980, however, the lawyers' efforts did get an indirect blessing from the Mormon church hierarchy. While aiming their unusual statement mostly at the government's plan to locate the MX missile system in southern Nevada and Utah, the faithful also interpreted the Christmas message from the First Presidency as covering the fallout issue. It read, in part: "Nuclear war, when unleashed on a scale for which the nations are preparing, spares no living thing within the perimeter of its initial destructive force, and sears and maims and kills wherever its pervasive cloud reaches." For the conservative church, even this oblique reference was a startling foray into temporal matters.[36]

Meanwhile, at a more private level the education of Dale Haralson continued during the fall of 1978. Haralson used the telephone extensively and crisscrossed the country to talk to experts on weapons tests, radiation, and cancer. He was absorbing basic knowledge, learning of pertinent studies and other documents, and judging who would make good witnesses. Haralson, who did not put much stock in the possibility of an administrative remedy or a negotiated settlement, was preparing for a trial.

One of the first people that Haralson contacted was Glyn Caldwell, who was working on the study of veterans. Haralson had gotten Caldwell's name from a *Parade* magazine article that mentioned the Smoky study. Caldwell, a longtime Public Health Service officer, was chief of the Cancer Branch of the Chronic Diseases Division of the Center for Environmental Health at the Center for Disease Control. The doctor was part of the federal health bureaucracy, which had been closely allied with the weapons testing bureaucracy for some time.[37]

Haralson broached the possibility of Caldwell being an expert witness for the plaintiffs, but Caldwell suggested that Haralson obtain independent experts. "I'm on the side of the government. You're the opponent," he said, and added, "I have already gotten burned once for having offered my services in a situation where it turned out the government would have some interest and gotten told that it was inappropriate and so, rather than putting in a lot of time letting you think I can really do something, that is why I put it out front. I'll help you as I can, but I could not appear in court." Haralson got the names of two cancer experts from Caldwell, one of whom was Joseph L. Lyon at the University of Utah. Caldwell alerted the lawyer to pertinent reports, sent him copies of some published studies, and suggested he contact the Nevada Operations Office of the test organization. Don't mention my name in connection with this, Caldwell repeatedly warned Haralson.[38]

Not all those who worked for the government reacted to Haralson's approach in the same manner. One outstanding anomaly was Harold Knapp, whose name Gofman had mentioned. Gofman had once served on a committee that had reviewed a controversial fallout study that Knapp had put together for the AEC. At one time Knapp was the maverick scientist within that tightly controlled organization. He was now working for the Institute for Defense Analyses, a Washington, D.C., think tank that did high-level nuclear warfare studies for the Pentagon. But Knapp had retained an interest in the health effects of the tests and was sympathetic to the victims, be they sheep or humans.

Knapp described his position as being "bizarre, supremely awkward, somewhat unprecedented, and potentially explosive." A congressional hearing was looming at the time, as was a second sheep trial, and Knapp told Haralson that he was operating in a "never-never land as to what the straightforward thing to do is." On the one hand, he was working with Department of Energy weapons officials in the course of his regular business. On the other, he did not know when a Department of Energy (DOE) attorney would subpoena his voluminous material on the subject.[39]

They met at the institute in October, and Haralson was held spellbound by the talkative Knapp for an hour and forty-five minutes. Knapp explained the workings of the AEC bureaucracy: "If you were interested in being promoted, of moving up within the bureaucracy, you followed policy, you did not hinder it." At the present time, Knapp said, he was working for the Joint Chiefs of Staff on nuclear weapons for the future. "So I have to be a little cautious that I don't do anything to be rude or alienate someone or sound wild," confided Knapp.[40]

It was a precarious position for the spirited scientist, yet Knapp managed to survive within the upper echelons of the defense bureaucracy. In describing his dilemma and how he had resolved it to a former AEC colleague, he said:

> My first inkling that something was up was when Mr. Haralson, the attorney from Tucson representing some of the persons who have filed claims, asked if he could come to Washington to talk to me. Next, KUTV of Salt Lake City wanted to send a camera crew to interview me for a documentary they have in progress. Next, the Public Broadcasting System people in New York [contacted me] for a documentary they are making. The thought of possibly somehow being publicly in the middle of 232 million dollars in claims against the Government isn't exactly my idea of a comfortable position. On the other hand, being a simpleminded New Hampshire man, I find it easy to empathize with the people in Utah and Nevada whose children drank milk and ate fresh vegetables during the years of heavy testing, having been explicitly assured by their Government that it was all right to do so, and subsequently developed cancer and died. So, after considerable puzzlement I came to the conclusion that if I'm asked, I'll simply try to explain as fairly as I can what I know, and with equal emphasis, what I don't know.[41]

Haralson next called on Charles Mays at the University of Utah. Mays recalled Knapp's difficulties within the AEC and some of his own experiences, among which was a warning that if he and Robert C. Pendleton, a colleague and an outspoken critic of the AEC in the 1960s, did not "knock off" their studies on thyroid cancer, federal funding for the lab would be cut back. The lab was now funded by the Department of Energy.

Mays had not forgotten that threat. "I have had my toes stomped on pretty hard and I would just really rather stay in the position where I could provide information but not be called upon to take sides," said Mays. The scientist agreed to act as an advisor, for a fee, and to channel Haralson information, some of which was quite useful. Haralson, recalling Gofman's warning, was guarded in his dealings with Mays and did not use him as a witness.[42]

Haralson first reached K. Z. Morgan by telephone and later visited him in Atlanta. To Haralson, Morgan was the ideal witness. Although somewhat hard of hearing and a bit absentminded, the lawyer found Morgan to have "total and complete objectivity when it comes to testifying in our case." Haralson considered Morgan's credentials to be impressive. At the time, he was in his early seventies, a professor at the Georgia Institute of Technology, and about to retire. He had helped to organize the profession of health physics in the 1930s and had worked on the Manhattan Project at the University of Chicago in the early 1940s. Morgan had been the first president of the Health Physics Society, the editor of its prestigious journal, and one of the founders of the International Radiation Protection Association. For some thirty years he had worked for the AEC as director of the Health Physics Division of Oak Ridge National Laboratory, in which capacity he had witnessed the tests in Nevada. Through the years Morgan had grown critical of the nuclear health establishment that he had once been so much a part of. Morgan was now beyond the criticism of his superiors and the reach of funding cuts.

He told the lawyer, "Very good. Well, as I said, I would like to do everything I can to help people or their survivors get some justice on unnecessary exposure and I feel very strongly that this was a case of unnecessary exposure to people." Over the next few years Morgan found Haralson to be "a very intelligent young man" who quickly mastered the complexities of health physics.[43]

The lawyer also had a fascinating talk with Dr. Ernest J. Sternglass, a professor of radiology at the University of Pittsburgh Medical School and the author of *Secret Fallout: Low-Level Radiation from Hiroshima to Three-Mile Island*. Sternglass had noted in the book a sharp decline in SAT scores in Utah that seemed to correspond to the fallout.[44] Referring to damages that might be awarded the claimants, Sternglass told Haralson, "That's the trouble, you see, the government really cannot open a breach in the wall voluntarily. At least, not until we beat their brains against the wall." Haralson found Sternglass, who said he would donate his time for such a worthy cause, too excitable to be a convincing witness for a sober-minded judge. Instead, he decided to use Arthur Tamplin, Gofman's colleague, who used an agent to collect his fees.[45]

Joseph Lyon, an epidemiologist at the University of Utah, was the kind of believable scientific witness that Haralson wanted, although he could get very little information from him at the start. Lyon was preparing an article for publication in the prestigious *New England Journal of Medicine* on the increased incidence of leukemia among children in the downwind region. The editors had strictures against the prepublication release of information, and an academic career could be on the line if those rules were violated. "The rule is now pretty well accepted that if one does that, one kisses goodbye to the opportunity to publish," Lyon explained to the lawyer. He said he had been getting "pushy" telephone calls from a *Washington Post* reporter. Haralson did not push Lyon, since he had a pretty good idea from Bill Curry of the *Post* and Chuck Mays, who had made some inquiries, as to what the study would say. Lyon was noncommittal about helping the plaintiffs and suggested that Haralson contact Mays. He warned him about Robert Pendleton. "He's very noisy," said Lyon. Haralson talked briefly with Pendleton, who was then in declining health.[46] Like a good journalist or lawyer, Haralson milked all possible sources to gain information.

Lyon had a hidden reason for not wanting to aid the plaintiffs. He had an application for more funding pending before the National Institutes of Health, and he did not want to offend the federal health bureaucracy. He notified NIH officials that the plaintiffs' lawyers would soon be requesting cancer statistics for southern Utah from 1950 to the present. "My opinion about that had better be left out of a letter of this nature," Lyon wrote his NIH contact.[47]

With what he knew about the Lyon study, Haralson called Alice Stewart, whom he had previously met in Morgan's Atlanta office, and asked her opinion of how that study might be applied to the claimants' situation.[48] Dr. Stewart, a British expert on the effects of low doses of radiation, had done a study on the effects of X rays on human fetuses. She believed that the low numbers of the Lyon study could be indicative of a number of things, including a statistical fluke. She apologized, and Haralson said that was one reason why he made such contacts by telephone, since written communications could be subpoenaed by the government. Because of Dr. Stewart's equivocation and the expense of bringing her from England to testify, Haralson decided against using her as a witness.[49]

By the middle of 1981, Haralson was winding up his search for credible expert witnesses and hard evidence. He wrote Nobel laureate Linus Pauling, who had taken on the government during the mid-1950s on the issue of global fallout, that "What I thought was a simple lawsuit for wrongful death has mushroomed into what is probably a piece of the most critical litigation in this century." Haralson visited Dr. Pauling in Palo Alto, California, and determined that, beyond the extensive media coverage that his testimony would generate, it would contribute little of legal value to the case.[50]

The process of discovery had now ended.

CHAPTER 2

THE BUREAUCRACY

AFTER THE lawyers acted, the government reacted. Like a ponderous creature that had slept overlong, the federal agencies involved in the weapons tests awoke by degrees to the fact that they faced a serious problem. The bureaucrats in the field were the first to feel the heat of the desert day.

One month after the first claims were filed, Malon E. Gates, manager of the Nevada Operations Office and the nearby test site, wrote a memorandum to his superiors in Washington. Gates, a retired army general, warned that "the initial wave of national interest" in the tests and their effects was about to break upon the Department of Energy. He cited newspaper stories, television interviews, Haralson's Freedom of Information Act requests, the filing of claims, and Udall's activities.

What could result from all these actions, Gates said, was a threat to the nuclear weapons program. "DOE programs and credibility could be significantly prejudiced by haphazard and piecemeal decision making in the interim," warned Gates, referring to a radiation-effects study being undertaken by the Department of Health, Education, and Welfare (HEW). He was most concerned about the interference of an outside agency in the affairs of DOE.

What was needed to combat this threat to continued testing was the coordination of counterefforts by the Nevada Operations Office. There should be a central repository of information. The advantage of such an archive would be to "prevent the dissemination of incomplete and misleading data." Gates also proposed "additional data-gathering activities and epidemiological studies of the population in the Southwest which may have been exposed." His office could serve as the contracting authority, but Gates warned that "care in selection of an organization must be exercised to assure objectivity and public acceptability."[1]

The emphasis from the start was on establishing a credible assessment effort by the same office that historically was, and still is, responsible for conducting the test program. Those two mutually contradictory roles were irreconcilable. They had plagued the believability of the test organization in the past, and they would do so again, regardless of whether the organization was known as the Atomic Energy Commission or the Department of Energy, the largest slice of whose budget was given over to nuclear warfare programs.[2]

The situation worsened by the end of November 1978. DOE officials in the front ranks in Nevada believed they were under siege. Inquiries from the public and the press and the demands of lawyers put the organization, which was used to operating with little public visibility, under an unaccustomed pressure. The memorandums flew back and forth between Las Vegas and Washington. Gates reported, "Each fresh media report causes the public to call NV seeking information on possible exposures and government programs for medical examination and care. The callers often ask for radiation sickness syndromes and medical advice." Gates also noted increased hostility from the public.[3]

Within the bureaucracy the pendulum began to swing away from the epidemiological studies proposed by Gates toward a dose assessment project that would be more useful in court. From this point on, the primary emphasis was on providing a legal defense for the government rather than a true assessment of the damage. A DOE attorney noted that the first set of interrogatories Haralson submitted "clearly reflects a good deal of work having been done over the past six, eight months, and clearly was put together by people who were familiar with the program." Another DOE official stated, "The principal difficulty for attorneys acting on behalf of DOE in court is the uncertainties in dose estimates, and the best dose estimates that can be obtained are essential for effective action of the DOE legal staff."

Such people as John A. Auxier, director of Industrial Safety and Applied Health Physics at Oak Ridge National Laboratory, and Charles Mays should be involved in such work. "Chuck Mays' involvement is important in that he is from Utah, has an excellent track record in this area, and has the confidence of the Governor of Utah," stated Walter H. Weyzen, manager of Human Health Studies Programs for DOE. Auxier was later named to the appropriate committee; Mays was not.[4]

While focusing mainly on a legal defense, Washington was not unmindful of the credibility gap. Gates sent a proposal for information gathering and dose assessment projects back to DOE headquarters, where the comment was made that the Nevada office "may be perceived as being too involved in testing" to head the projects. The suggestion was made to establish an advisory-review group of radiation and cancer experts from within and outside government circles "to improve our credibility."[5]

In March 1979, $250,000 was allocated for the gathering of documents related to the atmospheric tests and their health effects. As the documents began to surface from various depositories, it was determined that they had not been reviewed for twenty years. The public release of these documents and others that had previously been declassified was delayed in June by order of Major General J. K. Bratton, director of Military Applications within DOE. The general ordered that documents that had already been declassified be reviewed again "in light of recent events."[6] What he was referring to was rather embarrassing.

A student on leave from Harvard University who was doing research in support of an appeal filed by *Progressive* magazine, which had been legally blocked from publishing an article on how to make a hydrogen bomb, found on an open shelf of the library at the Los Alamos National Laboratory a declassified document that contained the secret. At a Senate hearing a few weeks later it was disclosed that yet another document that should have been classified was available to the public at the library.[7]

Other than this minimal start, which had been delayed, additional funding was slow in coming. General Gates exploded. In an unusually frank memorandum to his superiors that fall, Gates forcefully stated his frustrations:

> There are ironies associated with the dilemma encountered here by virtue of work required to answer interrogatories. Irony number one is that this work will have the effect of preparing the plaintiffs' case against the United States, while the work on OSRERP [the Off-Site Radiation Exposure Review Project] in preparing the defense against the suit is further delayed. Irony number two is that full funding for OSRERP is still denied me pending completion of a "staff report" by HQ, the apparent purpose of which is to justify to the Secretary my program for preparing the United States' defense against the suit. Clearly the fact that the suit has been filed is justification enough. Irony number three is that I am expected to flesh-out the outline of what I now consider an unneeded "staff paper," an effort that will further delay work on OSRERP and the interrogatories. Is anybody there?[8]

Full funding was not long in coming after this blast from the West. The DOE study, known as the Off-Site Radiation Exposure Review Project (OSRERP), was directed by DOE officials in the Nevada office, and its staff was drawn from the Los Alamos, Livermore, and other national laboratories and researchers under contract with DOE. Most had been employees of the old AEC. A peer review committee, named the Dose Assessment Advisory Group (DAAG), was established to monitor the project and give it credibility. The membership of the advisory group was heavily weighted toward those who had direct or indirect ties to government nuclear and health programs, such as John Auxier and Glyn Caldwell. Independence and credibility were lacking.[9]

The spring of 1979 seemed like a favorable time for the claimants' lawyers to attempt a settlement with the government. One of the government attorneys had written Udall, "I also want to assure you again that we are interested in working with you in every way possible in resolving these claims."[10] That certainly sounded hopeful.

In addition, the media was beginning to relate the story extensively at the national level, and most of the reports were sympathetic to the viewpoint of the claimants. Congress conducted hearings on the subject, which were also tilted toward the sheep ranchers and cancer victims. One startling revelation after another surfaced in the media and at the hearings.[11] The public image that was beginning to emerge was that of a conniving government versus an innocent, suffering people. Surely the government wanted to change that perception, the lawyers thought.

There were ample precedents for a settlement, which occurred in the majority of the larger, more complex civil cases confronting the Department of Justice. The advantage to the government, besides a change of image, was that a settlement would avoid a long, costly trial and appeals that could end in a judgment that was indecisive or, worse yet, that set an unwanted precedent. By avoiding a trial and the discovery phase that preceded it, the government would not have to make classified documents public, thus serving the needs of national security and also avoiding embarrassment. A settlement would also signal the government's compassion for its citizens. The disadvantages were that there would be a tacit admission concerning the harmful effects of radiation that might deter the test program, and a precedent would be set for compensation of an unknown number of future claims.

Rex Lee advocated settlement. He was in an unusual position at this time. Dean of the law school at Brigham Young University in Utah and a consultant to the plaintiffs' lawyers, Lee had been the assistant attorney general in charge of the Civil Division at the Department of Justice under President Gerald Ford, and he was soon to become the solicitor general in the Reagan administration, a position that would put him in charge of all government appeals.

Lee testified at the congressional hearings that when he was in the Justice Department, 95 percent of all the large cases were settled between the parties. "It is easier. It is quicker. It is fairer, and it saves both sides the costs. I think that is the approach that eventually ought to be taken." Not only was there a legal question here, said Lee, but there was also a moral one. "Federal policymakers decided to run some enormous risks. Innocent American citizens were involuntarily and unwittingly made the subjects of those risks and had thrust upon them the brunt of those risks." They should, Lee said, be compensated for the harm they had suffered.[12]

The plaintiffs' lawyers got together in Salt Lake City with Lee, who was a cousin of Udall's, for a strategy session. Should the suit be successful, there was talk of a $200-an-hour fee for the former assistant attorney general. Lee promised to approach a friend in the Justice Department and try and set up a meeting at which a settlement could be discussed. He told the attorneys, who included the western contingent and those in Udall's Washington office at the time, that the government could be expected to drag its heels for as long as possible. The federal government, Lee said, was conservative, more conservative than an insurance company, and had "very deep pockets."[13]

The first meeting between the lawyers on both sides was held on March 7 at the Department of Justice. Representing the government was William G. Schaffer, the deputy assistant attorney general for the Civil Division, and Bruce E. Titus, who had direct responsibility for the case. Schaffer explained that the Department of Justice was a client-oriented agency whose position was determined in this case by the Department of Energy and the Department of Defense. In this situation there was a moral as well as a legal responsibility, said Schaffer, who admitted that the government had made a mistake in the past. He said "the spirit was willing" to make a settlement, but first it was necessary to find out if there were any legal impediments. The government attorney added, "If the case is there legally, this behooves us to settle. If it is not there legally, we should still try to find a remedy." One such remedy would be to push a special relief bill through Congress. The attorneys representing the claimants at the meeting were amazed. Haralson noted, "Schaffer and Titus were clearly and totally in sympathy with us and anti-government in their attitude."

Chores were distributed at the end of the meeting. Titus was to check with DOE officials to see if they also favored settling the claims, and by the next meeting the claimants' lawyers were to submit a brief answering Schaffer's key question: "Do you see any legal impediments to a settlement?"[14]

As it turned out, the prospects for a settlement were never brighter than during the period between the two meetings. The attitude of the Department of Justice attorneys, who had touched bases with their clients, was less sympathetic at the second meeting in mid-April. The government attorneys now spoke less for themselves and more for their clients. From this point on, the attitude of the hard core of entrenched bureaucrats within the weapons establishment would stiffen on the way to the courthouse.

Schaffer was measured in both his tone and choice of words at the second meeting. His clients, he explained, were very cautious. They had a vested interest in this issue and wanted to study it some more. The opinions within DOE, said Schaffer, covered the whole range of possibilities, and it was impossible to predict what policy might emerge. DOE officials were most concerned about proof of cause and effect, or the lack of it.

The lawyers handed Schaffer the brief he had requested. The forty-page document referred to the Nevada Test Site and the downwind region as "the only radiation fallout laboratory involving humans ever established on this planet." In calmer tones, the brief went on to address the "decisive issue" of causation—the link between fallout and cancer. It stated, "Not only has the passage of time—and the Freedom of Information Act—lifted the veil of secrecy, but a new generation of outside radiation experts has emerged armed with epidemiological techniques which make it possible for them to use statistics to prove that certain levels of exposure will produce excesses of cancers in groups put at risk by the radiation-generating activities of the government."[15]

Schaffer closed the meeting by stating that he was not slamming the door on a settlement and that he was not personally being negative, but he was not sanguine about the attitude of those within the Department of Energy. He suggested that the lawyers meet with DOE officials, who were the principal clients.[16]

The meeting took place a few days later. At DOE the claimants' attorneys encountered stiff resistance to a settlement. The officials they met with were headed by Assistant Secretary for the Environment Ruth Clusen, whose advisors included a number of former AEC employees. They said they wanted to do the just thing, but it was also their duty to protect the public treasury. The causation issue was discussed. A DOE attorney, Joseph DiStephano, said the agency's "fundamental problems are more technical and scientific than legal." He added that a "substantial amount" of proof of causation was needed before any claims would be paid.[17]

The next obligatory stop on the bureaucratic treadmill was a meeting in mid-May with officials in the Department of Health, Education, and Welfare. This meeting was no more productive than the previous ones. The lawyers met with F. Peter Libassi, the general counsel of HEW. Libassi was the chairman of the Interagency Task Force on the Health Effects of Ionizing Radiation. The group's report was due out in June. Libassi said that any decision on the matter would be made by President Jimmy Carter.[18]

The president's attention was first drawn to the plight of the off-site civilians by the petition of the western representatives in Congress, most of whom were Republicans, and then, more forcefully, by Utah governor Scott Matheson, a Democrat who had lived in the downwind region during the period of atmospheric tests. Between newspaper and television stories; direct input from his constituents, the plaintiffs' attorneys, and Charles Mays; and the discovery of boxes of forgotten documents in the state archives, the governor had been made well aware of the issue. During a visit to Washington in November 1978, Matheson gave the president a memorandum that ended: "The Carter Administration has made a good record thus far on radiation health issues. It can avoid any criticism if it immediately orders a crash study of the cancer situation in the Utah-Nevada fallout area."[19]

When he got home, Matheson fired off letters to all the relevant cabinet officers, asking them to review their files for information and stating that he planned to disclose what he found in the state files, after consultation with them.[20] Meanwhile, President Carter directed Joseph A. Califano, Jr., the secretary of HEW, to reevaluate earlier studies on the incidence of cancer in Utah and consider the need for new ones.

Shortly after the presidential directive was issued, a Freedom of Information Act request by Bill Curry of the *Post* alerted Califano to the fact that there were important documents in his department's archives. Haralson had told Curry about a particular study. The reporter then requested the document and repaid the lawyer by telling him its contents. The resulting story, headlined "U.S. Ignored Atomic Test Leukemia Link," cited a Public Health Service (PHS) report by Clark W. Heath, Jr., that had long ago been suppressed by the AEC. Another PHS leukemia study was found in the Utah state archives at about the same time. Long-forgotten, classified, or suppressed information was popping up all over the place.[21]

When the Libassi report was released, it contained little about the plight of the offsite population, concentrating instead on more general issues pertaining to the health effects of low levels of radiation. The report simply noted that studies were being conducted and that the results would be announced at some future date. It did point out the difficulties in successfully suing the government in such cases and stated, "It is impossible to determine whether any given case of cancer was caused by radiation or by another factor." With DOE funding 78 percent of federally supported research on the biological effects of radiation, there was a publicly perceived conflict, the report timidly noted. It recommended that the NIH should take the lead in funding such research.[22]

Libassi was limited as chairman of the task force by the makeup of the group, which was drawn mostly from those agencies who either conducted or monitored the tests, or were the ultimate market for the product. He was less restrained in his congressional testimony. Concerning the downwind population, Libassi testified:

What seems to have been happening during this period of time was that in the early 1950s leading scientists believed that relatively low levels of radiation were, in fact, safe. During the late 1950s and 1960s there was increasing evidence that these levels of radiation were not safe. Despite this growing evidence, there seemed to be a reluctance or unwillingness to share with the American people the fact that there were growing questions about the safety of low levels of radiation. There seemed to be an unwillingness to address those issues, to pursue the research, and to disclose the information to the public. . . . The American people were not informed of the evidence that was gathered during the 1950s and 1960s on the health effects of radiation from these atmospheric nuclear tests. They simply were not informed of the uncertainties that we knew about at that time.[23]

The lawyers finally worked their way up to the White House, where Udall met in June with Frank White, an attorney on President Carter's staff who was keeping track of the radiation issue. White thought the president would be ready to make a decision in early 1980. Udall pointed out that it would be necessary to file suit soon in order to meet the legal deadline. White understood this and considered the filing of the lawsuit inevitable and perhaps necessary in order to move the bureaucracy toward a settlement.[24]

When given the information that there was no alternative but to file suit, Schaffer also agreed that such a move was necessary. He told Udall that the suit might give the Department of Justice more leverage and weaken the objections to a settlement that were coming from the hard-liners within DOE.[25]

The suit was filed on August 30, 1979. It was titled *Irene Allen v. The United States of America.* Irene Allen, who by alphabetical chance lent her name to the lawsuit, was a resident of Hurricane, Utah, a small community northeast of St. George. Twice widowed, Mrs. Allen was the mother of five children. Her first husband and two oldest boys had watched the tests from the roof of the local high school building. Her first husband died of leukemia in 1956, and her second husband died of cancer of the pancreas in 1978. Mrs. Allen worked as a secretary in the high school. She told a town meeting conducted by Senator Orrin Hatch shortly before the suit was filed: "I have really had quite a hard life, I feel, but I am not exactly blaming the government, I want you to know Senator Hatch. But I thought if my testimony could help in any way so this wouldn't happen again to any of the generations coming up after us, and I am really happy to be here this day to bear testimony to this."[26]

That same month the Carter administration embarked upon yet another radiation study. This one was undertaken by the Task Force on Compensation for Radiation-Related Illness, an outgrowth of a recommendation in the Libassi report.[27] Schaffer headed the task force, which President Carter formed to find a mechanism for compensating the downwind population. All the radiation studies Carter ordered were completed, but the president never took any decisive action on them. Schaffer later said that during darker moments he wondered whether the studies were ordered when no resolution of the issue was desired.[28]

At the first meeting of the task force, six of the thirteen people who attended were from the Departments of Energy and Defense. The remainder were from Justice, Health, and the Veterans Administration. From within the White House, the Office of Management and Budget and the Domestic Policy Staff were represented. At subsequent meetings, personnel from the Department of Labor and the Environmental Protection Agency joined the group.[29]

The problem Schaffer faced was how to mold a consensus from these diverse elements within the federal bureaucracy in light of their different involvements and perspectives on the test program and its health effects. There were power struggles between the different agencies. By coming in with a first draft that was overly sympathetic to the plaintiffs, Schaffer got the bureaucrats to settle down and work in a serious, concerted manner. "It was hard work and was almost at the level of a shouting match at times," said Schaffer.[30]

The report was sent to the White House on February 1, 1980. Newsmen were not able to get their hands on the document until more than a month later. The fifty-seven-page report was stamped "sensitive" and "official use only," but it contained no classified information.[31] What the report did contain, and what made it so sensitive, was the first admission by the federal government that some people had died from radiation-induced cancers as a result of the fallout from the test site. There was also a suggested mechanism for compensating the victims. The report represented the first time in thirty years of testing that all relevant agencies within the government looked at the tests and human health in a concerted fashion. These firsts were to become lasts, as the good intentions of the Carter administration remained just that, and the Reagan administration took over and firmly closed the door on any further admissions of guilt.

For however fleetingly, since it was never published and was soon forgotten, the report was a milestone. The admission read:

> It is well established that fallout exposed the population to ionizing radiation and that radiation exposure can increase the risk of many forms of cancer. Accepting the no-threshold hypothesis, we may reasonably assume that at least some additional cases of cancer in the downwind population resulted from atmospheric test fallout.[32]

After arriving at this startling conclusion, however, the task force then attempted to minimize it. The task force estimated that within a 250-mile radius of the test site, about 170,000 persons had been exposed to some degree of radiation. Stressing the uncertainty of dose estimates, the task force instead used monitoring data to estimate that 19 people had been exposed to more than 5 rems, 10,817 persons had been exposed to between 1 and 5 rems, and the remainder had received less than 1 rem.

Using dose-response statistics published in the controversial 1979 draft report of the National Academy of Science's Committee on the Biological Effects of Ionizing Radiation (the BEIR III report), the task force estimated that between 18 and 48 cancers above the expected number might occur, of which from 6 to 18 could be fatal. Applying an uncertainty factor of 2, the number of such cases could be as high as from 36 to 96, of which from 12 to 32 might be lethal. "Thus, from an overall public health perspective," the report stated, "the added risk to the downwind population from fallout was very small."[33]

Litigation as a means of settlement was rejected as being too costly and risky because of the chance of setting a precedent that could be adverse to the government's interests. "Further," the report noted, "given the nature of the illnesses, and the underlying national security reasons for the nuclear tests, resolution of these claims in a non-adversarial context is very much in the public interest." Again, this stated preference for a settlement was a first that quickly became a last.

The task force settled on an innovative approach to compensation. A "minimum probability requirement" would be established for each claimant. The idea was not unlike actuarial tables used by insurance companies. Using a number of factors that influence cancer induction from radiation, the probability of a relationship between fallout and a specific cancer could be calculated. Only those people who met a legislatively established minimum requirement would be compensated. The cost of such a program would be minimal because of the small number of potential beneficiaries. The report recommended that legislation be developed to implement this program.[34]

The plaintiffs' lawyers were not heartened by the Schaffer report. The feeble admission would cover only a few, and, needless to say, the lawyers would not share in this minimal settlement. Udall thought Schaffer had been captured by the hard-liners within the Department of Energy and the Office of Management and Budget and was "building a wall around us." Schaffer did not have much respect for Udall's legal abilities.[35]

Although the participating agencies, once they had a chance to review the report, had no significant reservations about its contents, the Schaffer report simply sat at the White House. Replying to a query from Senator Edward M. Kennedy seven months after the report had been delivered, Stuart E. Eizenstat, Assistant to the President for Domestic Affairs and Policy, said the compensation plan was being studied and that it would be several months before a decision was made. He said, "Designing a program which fairly and equitably provides compensation is exceedingly difficult. In our view, it would be productive to take the few additional months necessary to carefully examine all proposed solutions."[36]

Meetings continued to be held for a time among task force members, but then they tapered off. The complexities were still baffling.

The program was going to cost money, and the country was rapidly sliding into a recession. The hostages in Iran and reelection were higher priorities at the White House. Schaffer resigned his administrative position within the Department of Justice and took over as the chief trial attorney on the case. It was clear that the courtroom was where the action was going to be, although the dangers of an adverse judgment were well known.[37]

During three decades of testing at the Nevada site, there was only that one brief, albeit reluctant admission of responsibility by the government. The admission occurred in an administration that attempted to be open and populist in its orientation. The federal government soon reverted to the policy of placing national security above human health that had been firmly in place ever since the administration of President Harry S. Truman. The slight deviation from the historical norm was quickly corrected. In fact, it barely appeared in the log of the Ship of State.

After the Reagan administration took office in January 1981, the leadership of the country once again entered into a period of saber rattling reminiscent of the 1950s. There was talk of the evil empire and a program, dubbed Star Wars, that would depend heavily on the Nevada Test Site as a proving ground. Budgets for nuclear programs increased dramatically. The nuclear establishment, after a few years of public handwringing, retreated into the familiar posture of secrecy for the sake of the security of its citizens.

The government's position became one of absolute denial of any wrongdoing or negligent acts, or of any causal relationship between fallout and cancer in the offsite population. It vigorously pushed this position in the courtroom and in the legislative arena. The emphasis shifted from an attempt to come up with an equitable compensation plan at the administrative level of government to an expressed fear that any solution was a threat to the broad range of programs involving the use of radioactive materials, whether in the public or the private sector. Existing programs were to be protected but not the public who were the supposed beneficiaries of these programs.

Whereas lawyers had been the chief spokesmen and coordinators for the Carter administration, a general was the most visible spokesman for the views of the Reagan administration. The change in occupation and the predilection it implied were significant.

Lieutenant General Harry A. Griffith, director of the Defense Nuclear Agency, appeared twice before congressional committees and recited a litany of possible horrors should compensation be granted. Griffith testified that relief measures for the offsite population would result in a lowering of current radiation health standards, thus endangering the continued operation of academic research programs, medical and dental procedures, nuclear power plants, industrial radiology, nuclear ships, and the nuclear weapons program.

General Griffith also said that to encourage "the erroneous impression" that low levels of radiation were a health hazard would disrupt these programs in four ways. First, claims would be filed against the government and private industry that would place "a heavy burden" on these entities to disprove. Second, nuclear workers would become more difficult to recruit, and "the potential loss of manpower would stagnate the nuclear program." Third, compensation under existing health standards would result in those standards being lowered and "essential activities could be continued only with greatly increased cost while others could not be continued at all." And fourth, legislative and judicial recognition that low levels of radiation were hazardous "would increase the anxiety of the general public— itself an undesirable phenomenon—and thereby increase resistance to productive and necessary programs."[38]

What Griffith stated with undocumented certainty was not new. Echoes of these arguments could be found throughout the history of the testing program. The Defense Nuclear Agency (DNA) was the nuclear arm of the Department of Defense (DOD) and a direct descendant of the Manhattan Project. The DOD was the customer for what was conceived, manufactured, and tested by the AEC and its successor agency, the misnamed Department of Energy. Whatever the agency's name, the military, whether on active duty or retired, have always dominated the AEC and the DOE.

The sense of removal of these agencies from the mainstream of representative government was mirrored in their headquarters buildings. The redbrick DNA complex in the suburban Virginia countryside is nestled among woods and a golf course. The former AEC headquarters in the rural Maryland countryside, which now houses DOE's lower-level nuclear functions, is more a college campus than a government outpost, and the same could be said of the Los Alamos laboratory in New Mexico and the Livermore laboratory in California. The physical separation combined with the collegiate atmosphere bespeaks elitism.[39] DOE's nuclear inner sanctum is in the midst of downtown Washington but is set apart from other functions in the Forrestal Building by two security checkpoints and blue carpeting.

The office of Troy E. Wade II, Deputy Assistant Secretary for Defense Programs, had a low number, the guide pointed out, meaning that Wade was quite important. The office was suitably adorned with color photographs of Los Alamos, the test site, and missiles about to be launched.

Troy Wade had worked at the test site since 1958 and had risen to be the deputy manager of the Nevada Operations Office before coming to Washington with the Reagan administration. His views, like those of Griffith, with whom he was in agreement, were typical of the thinking that prevailed within this nuclear country club: The whole thing had been blown out of proportion; it was the lawyers, who had solicited the clients, that were causing all the problems; the government did not deliberately expose people to radiation, nor did the government control the wind.[40]

CHAPTER 3

CONGRESS

IF THE executive branch of government failed to find a solution, at least it could not be faulted for raising false hopes. The same could not be said of Congress, where public promises were rife but delivery on them was almost nonexistent. It could be fairly said that the attention the executive branch of government and Congress gave to the problem was in direct proportion to its public visibility.

Public concern and media attention over nuclear warfare and its consequences has been cyclic over the last forty-odd years. To deal consistently with such a subject at the level it deserved is not the American way, and it could lead to madness. The people who lived downwind from the Nevada Test Site got caught up in the whirlwind of the most recent cycle, only to be dropped at the end.

Three years after first testifying on the matter before a congressional committee, Governor Matheson told a similar inquiry in 1982: "As I appear before you today, I am far less optimistic and far more cynical about the outcome of these hearings than I was three years ago. I see little reason to believe that they, like their numerous predecessors, will result in answers the people of this state have sought for so many years."[1] Utah's Matheson had good reason for his skepticism. It was in 1957, at the height of public furor during a previous cycle of concern, that the first full congressional inquiry was conducted on the subject of fallout. The hearings, entitled "The Nature of Radioactive Fallout and Its Effects on Man," were carefully structured by the Joint Committee on Atomic Energy so that the AEC could tell its side of the story at a time of intense national concern over global fallout. Most of the four days were devoted to the issue of global fallout, local fallout in the Southwest being in addition to what was circulating worldwide. Through the late 1950s and early 1960s, the joint committee held more hearings, and then the issue seemingly disappeared when the tests went completely underground in 1963.

With the reemergence of the issue in the late 1970s, congressional interest once again quickened. The difference this time around was that the committee structure had changed. The responsibility of oversight was given to a number of committees instead of the one joint committee, which was sympathetic to the interests of the AEC. The membership of the different committees was not overly friendly toward the responsible bureaucrats, nor were they unduly protective of the interests of the weapons testing agency. The similarity was that little changed.

Congress was then faced with a host of cases involving toxins, and the apparent victims and their congressional supporters were crying out for compensation. The problem was that each case represented a narrow constituency, and the link between the poison and the harmed person was not directly traceable. Farm workers and pesticides, veterans and herbicides, veterans and fallout, civilians and fallout, neighbors of Love Canal, neighbors of Three Mile Island, asbestos miners and workers, uranium miners in the Southwest, occupants of homes in the West constructed from radioactive mine tailings, and Marshall Islanders who had been hit with fallout from a 1954 hydrogen bomb test in the Pacific were all clamoring for compensation. The chorus was deafening.

But Congress had a miserly record on providing compensation for such victims, regardless of whether the offending institution was in the private or public sector. Members of Congress were not scientists, the scientific community was split and suspect on these issues, and the institutions that were responsible for spreading the substances put up powerful defenses. There were costly exceptions. Congress appropriated nearly $1 million in 1964 for the Marshall Islanders and then added another $150 million in the 1980s for the inhabitants of the United States trust territory who had radiation-related cancers that were also familiar to the mainland population. Congress had passed a bill in 1969 compensating coal miners with black lung disease. By the late 1970s—when inflation, high interest rates, and budget deficits were rampant—the black lung program was costing more than $1 billion a year, and Congress was beginning to sour on that approach to solving these seemingly endless and intractable problems. Besides, there was always the courts.[2]

The goals of the 1979 congressional hearings were to find a method of compensation for the civilian population, to get DOE out of the business of radiation health research, and to gather all the documentation together in one place. Congress did a fairly good job on the last goal, since the three-volume hearing record consisted mostly of documents. But the first two goals were not accomplished. Congress made a brief attempt to dislodge the health research functions from DOE, then gave up. Legislation seeking compensation did not reach the floor of either house. Both the Carter and Reagan administrations consistently and adamantly opposed such legislation.

Senator Kennedy of Massachusetts and his staff spearheaded the hearing process in early 1979. Kennedy was then chairman of the Senate Committee on the Judiciary and the Subcommittee on Health and Scientific Research, and his party, the Democrats, controlled the Senate. The issue and its possible solutions were bipartisan from the start. Kennedy and Utah senator Orrin G. Hatch, a Republican, were political opposites, but they cooperated closely, as did their staffs. For Hatch, a junior member on both committees, it was a Utah issue that fit his committee assignments. He had not been in the state that long and faced charges of carpetbagging.[3]

Also participating in the series of hearings were the Senate Committee on Labor and Human Resources and the House Subcommittee on Oversight and Investigations. It was an impressive joint venture. When the committees met in Salt Lake City on April 19 for the first of four sessions, their members recited a litany of governmental wrongdoing, and solutions were promised in ringing tones. Senator Kennedy said, "Today's hearing is, in a very real sense, about the erosion of a people's trust and faith in their government." Senator Hatch said, "As should be obvious by what has been said this morning, today's hearing is no mere academic exercise."[4] But it was to prove to be just that.

There was heavy local and national media coverage the first day. Bill Curry of the *Washington Post* wrote that the charges made that day "represented perhaps the highest-level attacks ever made in a quarter-century of controversy about the health effects of nuclear testing in Nevada."[5] As the charges diminished, so did the cameras (or vice versa); and the committees got down to work.

The testimony and voluminous documents gathered at the hearings in Salt Lake City, Las Vegas, and Washington were compiled into three thick volumes and then synthesized into a forty-two-page report that contained the opinions of the House Subcommittee on Oversight and Investigations. The report, entitled *The Forgotten Guinea Pigs*, was issued on the thirty-fifth anniversary of the dropping of the first atomic bomb on Hiroshima. It was an indictment of times past and a call to "promptly compensate the victims of our mistakes," and it stated:

> Atmospheric testing of nuclear weapons in the 1950s and 1960s may well have been essential to secure the national defense. However, because the agency charged with developing nuclear weapons was more concerned with that goal than with its other mission of protecting the public from injury, the government totally failed to provide adequate protection for the residents of the area. There was sufficient information available from the beginning to suggest that if it was not possible to conduct the testing outside the continental United States, then the people living nearby needed protection. The necessary protection could have been provided by evacuating some of the people, but, at a minimum, the government owed the residents a duty to inform them of the precise time and place of each test and to instruct them as to what precautions should be taken. . . . The greatest irony of our atmospheric testing program is that the only victims of U.S. nuclear arms since World War II have been our own people.

As for compensation, the report pointed out that the responsibility lay with Congress, since a legal remedy was costly, time consuming, and unpredictable. The government could compensate the cancer victims on the basis of compassion. "Clearly," the report declared, "the government can compensate on the basis of legal liability, as a private person would be liable, or it may compensate on the basis of moral and equitable liability. The Subcommittee believes that sufficient evidence exists for the government to accept at least 'compassionate responsibility,' if not legal liability, for the injuries sustained as a result of the nuclear weapons testing program." The hazy legislative solution outlined in the report, however, did not clearly define eligible claimants.[6]

Like the Schaffer interagency task force report, which was the considered opinion of the executive branch of government at the time, the House subcommittee report came closest to being the definitive expression of Congress on the subject. Both agreed on the unsuitability of the judicial system, both agreed that something should be done, and both were unsuccessful in doing anything.

The futile search for a congressional solution was undertaken first by Kennedy and then by Hatch. In the fall of 1979 Kennedy—backed by ten other senators, including Hatch—introduced a catchall bill to compensate the owners of sheep and the human victims of the tests, along with uranium miners who had contracted lung cancer. For the downwind residents, the bill proposed a blanket solution that circumvented the thorny problem of causation. To constitute an "irrebutable presumption" that the injury was caused by fallout, all that was needed was residence in the area during the period of atmospheric tests and a radiation-induced cancer. Senator Kennedy pointed out that $7 million a year was being spent on studying the long-term effects on human health of the bombings on Hiroshima and Nagasaki, and the inhabitants of the Marshall Islands had twice been compensated for damages and illnesses.[7] Discussed by Hatch's staff but left out of the Kennedy bill was a provision for a cancer treatment center in southern Utah, which had first been proposed by an official at Southern Utah State College in Cedar City. Hatch favored such a facility but did not pursue it at this time.[8]

Hatch found himself in an unusual position. He had personally conducted a hearing in St. George and had been emotionally moved by the stories of human suffering. As one of the most conservative members of the Senate, Hatch recognized that the bill he backed was "admittedly a profligate, Great Society-type approach." On this issue the Utah senator found himself to the left of HEW officials, which, he said, was "a rare moment that is disorienting to all of us." Yet, Hatch asked, how else could the patient residents of southern Utah who remained good citizens be compensated? The senator wrote, "The issue is particularly agonizing for conservatives like myself. For years, we have advocated a strong national defense. We have instinctively tended to favor the tough-minded practical and economic case for nuclear power over the frequently emotional counter arguments. Now the contentions of those we dismissed have apparently returned to haunt us, in peculiarly horrible form."[9]

A hearing was conducted on the Kennedy bill in June 1980, but that was as far as it went in the Ninety-sixth Congress. The Carter administration opposed the bill because it was overly broad and too expensive. The Department of Justice noted, "The bill places no limitation on the amount of damages recoverable against the United States." The Department of Defense said that thousands of people with "normal" cancer deaths would be compensated under provisions of the bill but admitted that "some individuals living downwind from the Nevada Test Site may have experienced adverse health effects resulting from nuclear testing."[10] A similar bill that Congressman Gunn McKay of Utah introduced in the House never got to the hearing stage.

Following the 1980 elections, when the Republicans obtained a majority in the Senate, Hatch took over leadership on the issue. As the new chairman of the Senate Committee on Labor and Human Resources, Hatch introduced a bill in July 1981 that a conservative could support. It was somewhat similar to the Kennedy bill but set a limit on the amount of damages that could be awarded. The burden of proof was placed on the government to prove that it did not cause the cancers. Senators Kennedy, Hatch, Howard W. Cannon and Paul Laxalt of Nevada, and Dennis DeConcini of Arizona sought cosponsors, and they eventually came up with fifteen.[11] "We don't want to open the floodgates so that everyone in the country can claim they were victims of radiation, but at the same time we want to be sure that those who were affected can get help," said Hatch in a press release that accompanied the introduction of the bill.[12] The Utah senator was highly visible on fallout and cancer at a time when he was gearing up for reelection. His successful campaign would feature ads on this issue.

But the Reagan administration liked the Hatch bill no better than the Carter administration had liked the Kennedy bill. General Griffith, among others, testified against the bill. The most enlightening testimony was given by Victor P. Bond, an associate director of the Brookhaven National Laboratory, who was speaking for the National Council on Radiation Protection and Measurements. Bond suggested an innovative way to deal with the problem of causation. The degree of risk could be determined for an individual who had contracted cancer from radioactive fallout by arranging a number of factors in tables and then awarding damages in proportion to the percent of risk.[13] This probability approach bore a striking resemblance to the recommendation of the Schaffer interagency task force, but there was one major problem. The validity of the tables and their fairness had to be accepted by all parties. Given the divergence of opinion in this extremely controversial area, agreement would be quite difficult to obtain.

Senator Hatch asked the Departments of Energy and Defense how many people would be eligible for compensation under his bill. The Department of Energy estimated that the number would be less than 160 people if it were limited to the immediate downwind population and not applied nationally. The Department of Defense estimated that from 75 to 125 veterans, 300 to 1,000 civilians in the immediate area, and a total of from 2,000 to 5,000 civilians nationally would be eligible for compensation. Spokesmen from these two departments testified at a March 1982 hearing that they liked this approach better, but still could not support the bill.[14]

There was some opposition to and hesitancy about the bill within the Senate committees that had to pass on it. The concept was new, and cost estimates varied from $24 million to $28 billion. The bill was passed out of Hatch's committee, but it died in the Judiciary Committee, whose chairman was Senator Strom Thurmond, the conservative, defense-minded senator from South Carolina. The bill also got caught up in a personal feud between Hatch and Senator Howard M. Metzenbaum, an Ohio Democrat who was on the Judiciary Committee. Finally, the insurance industry lobbied against Hatch's bill, fearing that it would upset existing statistical standards.[15]

That Hatch, a committee chairman who also sat on the Judiciary Committee, could only get the most minuscule of bills passed on this issue, and then via the back door, was testimony to his ineffectiveness as a legislator, to the power of forces opposed to any change in how they had done business for years, to the concern about budget deficits, and to the renewed emphasis the Reagan administration placed on national defense and the Soviet threat. All that the bill had going for it was the death of a few Mormons in the Southwest and a record of deceit by the government.

At about this time, a clear voice spoke up in the readers' column of the *Salt Lake Tribune*. Darlene M. Phillips, a cancer victim who had lived downwind from the Nevada Test Site during the years of atmospheric testing, said she would gladly forego money from any compensation bill in exchange for care at a cancer treatment center that would also undertake research on the effects of fallout. "Damaged health," she said, "cannot be restored, the dead cannot be raised, and shattered families cannot be reunited this side of the grave. The notion of compensation is a game—a politician's game."[16]

In the waning weeks of 1982, Hatch attached an amendment to the Orphan Drug Act that would serve, at best, as a holding action. The act encouraged pharmaceutical companies to produce drugs for rare diseases which, without a subsidy program, would be uneconomical to develop. The rider Hatch attached to this popular bill called for the secretary of the Department of Health and Human Services to develop probability tables and to better define the factors that would determine the cost of compensation for the fallout claimants. But even this minor effort stirred up opposition within the Reagan administration. The Department of Justice strongly opposed the Hatch amendment, stating that such a new concept had not yet been officially reviewed by the national and international organizations responsible for radiation protection. Hatch shot back, "I strongly question the department's role as arbiter of science policy in this government."

The senator used his considerable influence at the White House and, after some hesitation, President Reagan signed the bill on January 4, 1983, despite his "grave reservations" about the amendment. The president pointed out that there was no consensus among experts that low levels of radiation caused cancer, yet the government was being asked to complete probability tables for just such levels.[17]

The president was incorrect. There was agreement that low levels of radiation caused cancer, but there was no unanimity within the scientific community on the number of cases caused by a specific low dose.[18] Just what the relationship was in terms of specific cases of cancer, it now became clear, would be determined by a federal court judge.

CHAPTER 4

LAWYERS
AND THE LAW

WHILE CONGRESS was attempting to find a solution, the lawyers on both sides continued preparations for the trial. There were potential witnesses to interview, documents to obtain, depositions to take, and pretrial motions to make in the Salt Lake City courtroom. The attorneys for the plaintiffs monitored the legislative process but never participated in it to any great extent. They never saw a companion bill working its way through the House and thus thought that there was little hope for the Hatch bills. Generally, these bills gave the lawyers less money than they would receive if the litigation were successful.

On the Sunday after Thanksgiving in 1981, Haralson felt a swelling on his throat while shaving. Being extremely cancer conscious, the attorney made an appointment the next day with a doctor. The swelling was diagnosed as throat cancer—a squamous cell carcinoma —with the primary site at the base of the tongue. For a trial lawyer about to argue the biggest case of his career, no other location could have been more crippling.[1]

Haralson asked his doctor about the prognosis. Not good, he was told. He asked if it was the type of cancer that would rapidly spread throughout his body. Yes, said the doctor. But another doctor was much more optimistic. He foresaw a complete cure. Like most of the plaintiffs, Haralson eventually went through surgery, chemotherapy, and radiation treatments. He could now identify completely with his clients.

The attorney resumed his law practice in January and began

preparations for the Allen trial. But in late February he entered the hospital for major surgery, to be followed by two months of radiation therapy, one of whose side effects would be a temporary loss of voice. It was obvious that Haralson was not going to be ready for the April trial date, so he asked for a continuance. Judge Bruce S. Jenkins, who had been selected by lot to hear the case, set a September 1982 trial date.

After six weeks spent recuperating from surgery in the hospital, Haralson began extended radiation treatments. Five days a week he went to the hospital. He was in and out of the door in fifteen minutes, but during that time he received a dose of 5,400 rads. Haralson felt nothing. He just heard a slight buzz. Haralson was aware that radiation can kill—indeed, he believed that some of his clients had died from exposure to just a few rads—but he was desperately hoping that it could also cure. The lawyer prayed; and his family, friends, and others prayed for him. He was confident that he was going to recover, and he eventually did. Haralson realized, however, that he was not going to recover in time to complete the pretrial discovery process before the September trial date, and he would not be strong enough to withstand the rigors of what was expected to be a two- to four-month trial. Reluctantly, very reluctantly, Haralson came to the conclusion that another attorney would have to take over as lead counsel for the plaintiffs.

The government, too, was faced with a change of chief counsels at this time. Schaffer left the Department of Justice to go into private practice, and Bruce Titus, who replaced him, similarly departed in April. Henry A. Gill, Jr., a Department of Energy attorney, became the chief trial lawyer for the government. Gill, who came from Boston, had been a naval bombardier on nuclear-armed jets before taking up the law. He believed in a strong national defense.[2] At about the same time that Gill took over, the government lost some other attorneys who had been working on the case. Gill asked for a continuance beyond the September trial date so as to be better prepared to present the defendant's case.

From the standpoint of strategy, Haralson did not want a continuance. The plaintiffs' case was as ready as any such complex litigation could be, and he wanted to take advantage of the government being shorthanded. Haralson contacted three high-powered trial attorneys, all of whom were interested in the case but wanted a postponement in the trial date for the same reason that Gill did. He searched further. One of his partners was a friend of Ralph Hunsaker, a Phoenix lawyer, and suggested that he might be available.[3] Haralson knew of Hunsaker, a native Arizonan and former quarterback for the University of Arizona. He was competent, thorough, and committed to the projects that he undertook. Hunsaker was a Mormon and, thought Haralson, would have a strong empathy and common bond with his clients. Also, the Phoenix lawyer was opposed to a continuance. Haralson asked him if he wanted to head the team, and Hunsaker said yes.[4]

After Hunsaker took over the case in June, Haralson stepped back into the role of an active consultant. He gave advice and handled some of the witnesses and participated in the opening and closing arguments at the trial. It was Haralson's strategy and the case that he had so carefully prepared over the years that Hunsaker tirelessly implemented. The financial arrangement called for Haralson to split his 40 percent share of the fee evenly, if there was an award, with Hunsaker's firm. Haralson wrote his new colleague that the trial would be "the most intriguing and, hopefully, the most rewarding litigation that either of us will ever see."[5]

The government's motion for a continuance was denied. The trial preparations became more hectic during the summer months, when scores of last-minute depositions the government had requested were taken all over the country. Hunsaker thought these requests were deliberate harassment by the government, an attempt to spread the plaintiffs' resources thin at a time when they needed to concentrate them on preparing their case. Gill later said this was not so.[6]

During the summer, the plaintiffs' lawyers made two final attempts to negotiate a settlement. Government attorneys either rebuffed these offers or did not respond to them. They were confident of their case and thought that if they did not win at the trial court level on the causation issue, the judge's opinion would be reversed at the appellate level on the discretionary function exemption.[7]

Udall made the first attempt at a settlement in midsummer. Citing the congressional hearings and studies favorable to the plaintiffs, he proposed to Assistant Attorney General J. Paul McGrath, who was in charge of the Civil Division, that there be a two-phase negotiation process that would first be a test of the leukemia and thyroid cancers and then of the other types of solid cancer cases. There was more evidence linking leukemia and thyroid cancers with fallout than other cancers. Udall threw out a figure of $123 million for settlement of the 1,200 claims. "We also wish to candidly state that we are opposed to any kind of protracted negotiations which could lead to a further postponement of the September trial," said Udall.[8]

Udall met twice with McGrath. The government lawyers were not impressed with Udall's one-and-a-half-page memorandum or his oral arguments. His arguments were too thin and had too much fluff. "He seemed to have the idea that if he knocked on the right door, had gone high enough, and had created enough adverse publicity, the legal and factual issues would just go away. He didn't know what he was doing," said Jeffrey Axelrad, the director of the torts branch.

The $123 million was out of the question. Besides, the government lawyers were convinced that they had a strong case.[9] But they threw out one possible bone. In an informal discussion with Udall, Gill floated the suggestion that the government might be willing to fund a cancer treatment center in Utah. After all, the Marshall Islanders had gotten free treatment. Such a solution seemingly skirted the government's fear of setting precedents. It would also take care of present and future generations. Udall said it would make a nice icing on the cake. But it was only Gill's idea—he had not cleared it with his superiors—and it went no further than that.[10]

The two opposing groups of lawyers, who were irreconcilably locked into their respective positions, met in mid-August in Salt Lake City. Those representing the plaintiffs, who had grown to 1,192 by this time, suggested a settlement. The amount was clearly negotiable. No counteroffer was made by the government, and the last effort aimed at negotiating a settlement died in silence. The trial that nobody wanted and that nobody could win was about to begin.

By default, an exceedingly complex issue involving national defense, public health, individual suffering, scientific proof of causation, and an historical search for negligence devolved upon a single federal judge. The judge and the inquiry he conducted were constrained by the law—in this case the Federal Tort Claims Act. The history of the law demonstrated the government's reluctance to be sued and the difficulty in obtaining a settlement. It was as if the government gave with one hand, then took away with the other.[11]

Prior to passage of the act in 1946, Congress had been deluged with individual relief bills, the alternative to recovering losses if the government could not be sued. What the federal government invoked to avoid lawsuits was the doctrine of sovereign immunity, a holdover from the days of absolute monarchs, when the motto was "The King can do no wrong." In 1832 Congressman John Quincy Adams complained that the special relief bills took up half of Congress's time. With the growth of the federal bureaucracy and the use of government motor vehicles by employees in this century, the situation worsened. The special bills clogged the claims committees. Between 1921 and 1946, over thirty bills attempting to remedy the situation were submitted to Congress. According to Lester S. Jayson, a legal scholar, those who opposed the remedies "feared that the government would be made the victim of ambulance chasing, exaggerated or fraudulent claims, and emotionalism in verdicts."[12]

Then in 1945 a military aircraft crashed into the Empire State Building in New York City. The doctrine of sovereign immunity barred any recovery of damages for the relatives of the dead and injured. The tort claims act, with twelve categories of exclusions, became law the following year and was made retroactive to cover the airplane crash. But the federal government did not lightly surrender the doctrine of sovereign immunity. Cases were to be decided by a judge, not a jury, which was likely to be more compassionate toward victims, and the exclusions made successful litigation difficult.

Two of the categories of exclusion from legal action—the statute of limitations and the discretionary function clause—seemed applicable to the Allen case. Government lawyers pressed their arguments on these legal points before, during, and at the conclusion of the trial. Judge Jenkins ruled against them each time. It was on these matters of law that the government hoped to win the case at the appeals court level.

The two-year statute of limitations clause seemingly extended from when the claimant discovered the negligent act, although some opinions have specified the date of death or diagnosis, which was the government's contention in the Allen case.[13] All of the plaintiffs seemed to fall within the two-year limit if it ran from the date of discovery. Almost all would be excluded if the clock started ticking from the date of death or diagnosis. The lawyers warned the plaintiffs to be careful about what they said to the news media, since they might jeopardize their case on the basis of the statute of limitations. Haralson wrote his clients in 1980: "I am concerned that some of my clients may have heard so much publicity in the past two years that they may conclude that they have known 'all along' that the radiation caused the illness. As you will recall, the Atomic Energy Commission went to great expense to assure the public that there was no connection between radiation and cancer or leukemia." He reminded his clients that the first news stories had surfaced in 1977, and he had first come to St. George in 1978. Haralson said those dates would help the claimants determine when they became aware of the illness, and he urged them to keep the letter confidential.[14]

Of the two categories of exclusion, the government lawyers pushed the discretionary function argument harder. The purpose of this provision of the act was to exempt policy decisions from legal action. In other words, it allowed the government to govern. The question in the Allen case was: Where did high-level policy-making and lower-level planning leave off and standard operating procedures begin? In the definitive textbook on the law of torts, William L. Prosser said there were cases where courts had held that operational decisions in the field fell under the "discretionary" umbrella. He added, rather ambiguously, that "there have been as many cases, however, which have held that such 'operational' exercises of discretion are not those contemplated by the Act; and this seems definitely indicated as the position of the Supreme Court."[15]

Jayson wrote in *Handling Federal Tort Claims*, a standard reference work on the subject, that the act had been designed to lift the burden off Congress and to right wrongs; and hence should be construed liberally. "The liberal approach, or at least rejection of the doctrine of narrow construction, is clearly the settled view of the

Supreme Court today," wrote Jayson in the book that was published two years before the court changed direction on the issue. Discretionary functions are the regulatory acts of agencies, policy determinations by executives, decisions of administrators, their plans and designs, "and analogous conduct and activities of federal agencies and officials requiring the exercise of judgment and discretion." [16]

To win their case, the plaintiffs had to prove negligence, causation, and damage. Negligence is defined as "conduct which falls below a standard established by the law for the protection of others against unreasonable risk or harm." [17] Causation involves linking a specific case of cancer to the fallout. Damage means actual monetary loss, such as income and medical expenses. Loss of love, companionship, comfort, and guidance are less tangible but nevertheless real bases for claims for damage. The burden of proof lies with the plaintiffs. It takes a preponderance of evidence, not absolute certainty, to prove negligence, causation, and damage.

That was the beauty of the law as opposed to science, most of whose practitioners were loath to take a position in this matter until there was certainty. The law recognized a degree of uncertainty. As Supreme Court Justice Benjamin N. Cardozo wrote in his famous treatise on how a judge decides a case: "As the years have gone by, and as I have reflected more and more upon the nature of the judicial process, I have become reconciled to the uncertainty, because I have grown to see it as inevitable." [18]

The Allen case fell into a category of lawsuit that has become more prevalent in recent years. Increasingly, the seemingly intangible cause-and-effect relationships between nontraceable toxic substances and diseases—such as between dioxins and cancer, and between dust particles and black lung disease—have been probed not only in the courts but also in the media, the scientific community, and the legislative and administrative arenas. The classic example of such an issue is the link between cigarette smoking and lung cancer.

CHAPTER 5

THE SITE

PROMPTLY AT 9:30 A.M., as would be his habit for the next two months, Judge Bruce S. Jenkins entered the oak-paneled courtroom on the second floor of the United States Post Office and Court House fronting on Main Street. The squat granite building of the neoclassical revival style was built shortly after Utah became a state in 1896. With such a building the federal government sought to establish a visible presence in Utah following bitter conflicts between Mormons and gentiles and federal and territorial governments. Within the confines of the solid gray structure an inquiry into a different type of federal presence was about to begin. Folding chairs supplemented the wooden benches in the spectators' section of the courtroom, and all were filled on the overcast morning of September 14, 1982.[1]

A tall, gaunt man with a thatch of black hair who bore a certain resemblance to Abraham Lincoln, Judge Jenkins came to the federal judiciary by following the twin paths of politics and the law. After graduating from the University of Utah law school, Jenkins ran unsuccessfully for city judge, member of the Utah House of Representatives, mayor of Salt Lake City, and U.S. Representative. Jenkins, a Kennedy Democrat in a conservative Republican state, had the additional disadvantage of being noticeably shy for a budding politician. He was appointed, then elected, to the state senate and rose quickly to head that legislative body. Before being named to the bench by President Carter, Jenkins had served as a referee in the federal bankruptcy court for six years. A native of Salt Lake City, Jenkins is a Mormon "by inheritance and inclination," but the judge is one of those rare individuals who transcend their roots. He reads voraciously and has an inquiring, lively mind.

By the time the Allen trial was over, Jenkins had come to be respected by the lawyers on both sides for his fairness and quick grasp of the complex issues involved in the case. To the lawyers who appeared before him, Jenkins stressed thoroughness of preparation. He was impatient with those attorneys at the trial who were not well prepared. Great care, he said, should be taken to distinguish substance from symbol. Jenkins showed little interest in impassioned arguments, pejorative wording, and rhetoric for its own sake. Clarity, lucidity, simplicity, thoroughness, and precision of presentation were all qualities that Jenkins admired.[2]

Before all the spectators had time to rise, the judge was seated. The case had been before him in its pretrial status for three years. Without further comment he said, "Let's go ahead in the matter of Irene H. Allen and others versus the United States of America."

Following the opening statements, the parade of witnesses began. The witnesses sat unnoticed among the spectators with the burden of their memories for company. Then, when summoned, they rose, shed their anonymity, and had their brief moment in the judicial spotlight. They fell into one of three distinct categories. There were those who oversaw weapons production or planned, conducted, or explained the tests to the public. Then there were the victims, either those who had cancer or the relatives of the dead. Finally, there were the scientists, who sought to prove or disprove a link between the tests and the victims. In this manner, the possible cause was followed by the possible effect, and the proof of a link, or absence of it, came last.[3]

The witnesses of historical consequence were old men who had become minor celebrities by virtue of the fact that they had participated in the birth of the nuclear age. Many who took part in the selection of the continental site and the conduct of atmospheric tests were dead, but their words and deeds were recorded in numerous documents submitted as evidence at the trial or found elsewhere in various archives. To have so great a time span between the crime and the trial was unusual for a legal proceeding.

The story began with Norris E. Bradbury, who had been there at the beginning. Bradbury was one of the cadre of high-ranking bureaucrats who worked on the weapons program for many years in relative obscurity and who far outlasted the nominal policymakers who came and went through the revolving doors at the highest levels of government. J. Robert Oppenheimer was the director of the Los Alamos laboratory for its first two years, and Bradbury was director from 1945 to 1970. During the early years of the Bradbury administration, the impetus for a continental test site arose. Once the Nevada site was chosen, it was mostly Los Alamos personnel who directed the tests and evaluated the results from the standpoint of weapons effectiveness and offsite radiation hazards during the years of atmospheric testing.

Like many others at Los Alamos, Bradbury was a product of the University of California at Berkeley, where he ranked among the top 10 percent of the physics graduate students in 1932. One of his professors, Ernest O. Lawrence, who was to win a Nobel prize for inventing the cyclotron, said Bradbury "has a good personality and presence, and is well liked by everybody." After receiving his doctoral degree at Berkeley, Bradbury quickly rose to a full professorship at Stanford University and, as a naval officer during World War II, worked on ordnance research until he was assigned to Los Alamos in 1944. Bradbury was the associate leader of the Explosives Division under George B. Kistiakowsky and helped develop the spherical implosion method that detonated the first atomic device.[4]

On Friday, July 13, 1945, Kistiakowsky and Bradbury drove "The Gadget" over the bumpy roads from Los Alamos to the Trinity test site near Alamogordo, New Mexico, where Bradbury assembled the nonnuclear components of the device in a tent under the tower. Shortly before the test, the young scientist noted in his log, "Look for rabbits' feet and four-leaf clovers." Ever the pragmatist, Bradbury said of the first nuclear detonation: "My immediate reaction on seeing Trinity was, 'It worked!' "[5]

Following the Trinity test, some one hundred head of cattle ten to fifteen miles downwind from the blast received burns from beta particles on their backs and necks. The AEC observed these cattle for a number of years but reported no long-term effects from this first fallout episode other than the graying of hair on the burned areas. Edward Teller, who has played a key role in the nation's nuclear program from the early days of Los Alamos to the Star Wars concept of the Reagan administration, later wrote that the test "gave us reason for grave concern about radioactive fallout in the vicinity of explosions." If there was such concern, it was never clearly voiced. AEC officials would later dismiss similar symptoms in sheep as indicative of malnutrition.[6]

After the test, Bradbury trained the teams that assembled the two weapons that were dropped on Japan the next month. He had no moral reservations about working on the weapons program after the war. "Look," said Bradbury, "this country's got to do it; we've got to go on from here. We haven't really scratched the surface of this business. You better be ahead if you're going to live in this world."[7]

It was Oppenheimer who selected the young naval commander to succeed him as director of the laboratory. When Oppenheimer's loyalty was being questioned in 1954, Bradbury, who had known and been friends with the famous physicist since Oppenheimer's graduate school days, testified at the AEC security board hearings that "here is a man who is completely and unequivocally loyal to the best interests of this country."[8]

When World War II ended, the quasi-military encampment on the isolated volcanic plateau in northern New Mexico went into a period of decline. But the Los Alamos Laboratory, into whose title the word "scientific" was inserted in 1947, was rescued from a premature demise by the Cold War and Norris Bradbury. The director was convinced that it was in the national interest that Los Alamos remain open in the immediate postwar years. He said, "We had, to put it bluntly, lousy bombs." The reliability, efficiency, versatility, size, and weight of the weapons needed to be improved.[9] There were, in fact, no bombs that were immediately usable, and components for only seven complete nuclear weapons existed in the national stockpile. When President Truman was apprised of this fact by the AEC commissioners on April 3, 1947, he was visibly shocked.[10] Within a month a commitment was made to expand the weapons program, and by the following year Los Alamos, which was once again humming with activity, had produced fifty weapons for the stockpile. The decision to build the hydrogen bomb in 1950 further increased activity at the laboratory.

Bradbury had a firm hand in the laboratory's rejuvenation. Ernest Lawrence said, "He astonished all those who had prophesized that the effectiveness of the laboratory would disappear with its most famous scientists, and that the morale in the isolated community would sink out of sight." Along with more prosaic management techniques, Bradbury brought jazz and professional wrestling to "The Hill" to boost spirits. A psychiatrist was also summoned to treat neuroses.[11]

Part college campus, part military headquarters, and part industrial complex, Los Alamos in the early Bradbury years pretty much went its own way with all the funding it needed. Other than determining salaries and other minor housekeeping chores, the University of California, which operated the laboratory for the AEC, had no control over the programs. The university was not even consulted about Bradbury's appointment.[12] The laboratory director explained how he dealt with the university's governing board. He said:

> Quite frankly, if you had to go see the regents, you would whomp up a nice show-and-tell sort of thing. The object was to make it look good, make them pleased with their operation. And you would whomp up just enough things that were not very classified, so they could feel they knew something without anything being harmed if they talked about it.[13]

Except for major policy decisions, such as whether to develop a thermonuclear weapon, Los Alamos—not the personnel at the AEC headquarters in Washington—formulated specific policy. "Policy flowed from here to Washington, and then came back to us as official," said Bradbury. He added, "You couldn't have policy come out of a bunch of bureaucrats in Washington, except in the most general terms." One of Bradbury's colleagues, Richard D. Baker, recalled, "Norris decided even before the [Atomic Energy] Commission was formed what he thought the Laboratory should do, and when the Commission was formed, putting it bluntly, he sort of told them what the Lab would do." [14]

Bradbury's life was The Laboratory. At a 1980 oldtimers' reunion for those who had worked at Los Alamos during World War II, Bradbury, who continued to live in Los Alamos in retirement, told his "fellow alumni" that "It tied me to it for thirty-five years. It is my home. I love it. It is an emotional experience to be here." One of the few areas the public is allowed to visit on The Hill is the Norris E. Bradbury Science Hall, a museum of nuclear artifacts dedicated to "his leadership, vision and devotion to the nation's interest." [15]

A short, spry man whose manner was terse and emphatic, Bradbury was seventy-three when he testified at the trial. A division leader under Bradbury once said, "You know, he talks like a machine gun. As soon as he gets what he wants, he whirls off and goes off in another direction." The court reporter had difficulty keeping up with Bradbury, who said, "Without testing, the program would have come to a complete stop. There is no way to guess how a bomb would perform if you haven't tested it. Unless you can test, you might as well go back to high explosives." Of the effects of fallout, Bradbury remarked: "Sure there are a few people with leukemia. More people get killed in automobile accidents every hour than ever will die of leukemia." [16]

The first postwar atomic tests took place in the waters off Bikini Atoll in the Marshall Islands, about two-thirds of the way across the Pacific Ocean, after anchorages in the Caribbean Sea and off Ecuador's Galapagos Islands had been considered and dropped. A bomb was released from an airplane and another was detonated underwater in the summer of 1946 to determine the effect of nuclear weapons on naval vessels. The most noteworthy feature of the second test, the commander of the joint task force reported, was "the extent and persistence of the radioactive contamination of the target ships." [17] The weapons for the tests were made at Los Alamos, and a large number of people from the laboratory participated in Operation Crossroads. Of the 9,000-mile journey from the mainland to the test site, Bradbury said, "It was extremely difficult, time consuming, expensive."[18]

The next year Bradbury suggested the possibility of a continental test site, but the proposal was not pushed very hard. At a June meeting in the Cabinet Room at the White House, AEC chairman David E. Lilienthal outlined the need for tests in 1948 for President Truman. Secretary of War Robert P. Patterson asked why the tests could not be conducted at the previously used Trinity site in New Mexico. General Dwight D. Eisenhower, Chief of Staff of the Army, opposed use of any continental site because he thought the tests would spread fear among the civilian population. Lilienthal also preferred the Pacific site. Truman authorized the test series that day and left the details to Patterson and Lilienthal to work out. Project Sandstone, which consisted of three tests, was conducted at Eniwetok Atoll in the Marshall Islands in the spring of 1948.[19]

The Cold War intensified greatly over the next two years. The Berlin blockade, the loss of China and Czechoslovakia to the Communists, the discovery of atomic spies, the growing concern about Communists within the government, the Soviet detonation of an atomic bomb, and the start of the Korean conflict all resulted in a greater impetus to increase the nation's nuclear arsenal under more secure conditions.

There was also the continued inconvenience of the Pacific testing grounds. The availability of transportation, the time-consuming journey for personnel and equipment, cost, and the logistics once there were all factors working against the Pacific site. Besides, some scientists and technicians at Los Alamos were apprehensive about the long flight over water. "The basic reason for activating a continental test site was to provide Los Alamos with a backyard laboratory for diagnostic tests vital to the weapons development program," stated a 1953 AEC report on the future of the test site.[20]

Bradbury and the military also had reservations about the security of the Pacific site. There were fears that unspecified "foreign powers," meaning the Soviets, might observe and even interfere with the tests. A continental site would be more secure. Further, predicting weather in the tropics was almost impossible. A more reliable weather pattern and the ability to predict changes were needed to lessen the dangers of fallout. Although no specific site was named, the Atlantic coast was mentioned in a memorandum from Rear Admiral William S. Parsons, the deputy commander of the task force that conducted the 1948 tests. Parsons noted, "There is no question that there will be difficult local and general public relations problems."[21]

This was the first mention of public relations in reference to a continental site, but it was a term that was used frequently thereafter as the government searched for a site, announced its selection, and, once the atmospheric tests got underway, sought to deflect perceived threats to the program and attempted to make the tests palatable. The term was used interchangeably as a palliative and a cure.

Admiral Parsons, who had been in charge of making the scientists' invention combat effective at Los Alamos, was the leading military expert on atomic weapons. Like others in the military in the immediate postwar years, Parsons wanted to counter the public's perception of these weapons as awesome machines of mass destruction, because such an attitude might hinder the military capability to test or use atomic weapons. To Parsons the "reservoir of unjustified fear" had to be eradicated and the public shown that it was "just another bomb" by conducting atomic tests on the mainland. A continental test site, said Parsons, would "lay the ghost of an all-pervading lethal radioactive cloud which can only be evaded by people on ships, airplanes, and sandspits in the Marshall Islands."[22]

Parsons' memo was presented to the AEC commissioners in early September 1948 as background material for a request for a preliminary survey of possible sites. The concurring AEC staff report noted, "It is believed possible that a properly conducted public information program stressing radiological safety factors as a result of prevailing winds could overcome adverse public reaction. Irrespective of location, the Commission could expect an increase in legal claims arising from real or fancied damage as a result of test operations." That prediction was to come true in spades thirty years later.[23]

At the September 16 commission meeting Chairman Lilienthal,

the former chairman of the Tennessee Valley Authority, expressed "strong doubts" about testing within the United States and said that if it was known that such an activity was even being considered, there would be "unfortunate public repercussions." The commission, however, agreeing with logistical and cost arguments, unanimously authorized a secret study. The study was to be conducted by the military with "a minimum of field activities likely to arouse curiosity." Lilienthal emphasized the "very tentative" nature of the commitment.[24]

The resulting fifty-seven-page study, code-named Project Nutmeg, was completed in January 1949. The report discussed the effect of radioactive fallout on photographic film, plants, animals, and fish —but not humans. It also contained a summation of what was known about the geographical extent of fallout at the time. Fission products from the 1948 tests were detected halfway around the world, but no prohibitive levels of radioactivity were found three hundred miles beyond the test site, a distance that easily encompassed the downwind communities.

The report contained a cursory description of some possible sites in the Southwest and on the Atlantic Coast. It focused most closely on the North Carolina coastline between Cape Hatteras and Cape Fear, where it was thought that the Gulf Stream would dilute the fallout by dispersing it throughout the Atlantic Ocean. There was no field inspection of the coastline, in accordance with AEC orders, and the basis for the analysis was aerial photographs and nautical charts. Again, public relations was singled out as the single most important factor in the delicate continental equation. Except for a quick engineering survey of the Nevada site in the summer of 1950, this military study conducted by a Navy captain was the only evaluation of a continental site, or sites, before its selection.[25]

When AEC personnel flew over the North Carolina coast following the submission of the Nutmeg report, they determined that almost all usable sites were inhabited and that it would be too costly to fill in the wetlands. The local residents would have to be either temporarily evacuated each time there was a test or permanently relocated. Coastal ship and air traffic would be disrupted during periods of testing. For these reasons, the location was unacceptable.

As a result, in March 1949, during a lull between external threats to the country, the AEC notified the military that "except for test activities during a national emergency, a continental site is not desirable." At the time, President Truman also felt strongly that the tests should not be conducted within the United States.[26]

The situation changed drastically within the next sixteen months. The Russians unexpectedly detonated a nuclear device in August 1949. In January 1950 President Truman ordered the development of the hydrogen bomb, and the following June the North Koreans invaded South Korea. The search was on again for a test site on home ground. The military had previously taken the initiative; now the scientists at Los Alamos would push in a determined manner for a continental site.

The laboratory was under intense pressure to develop the hydrogen bomb. A site was needed both to test the fission devices that would detonate the fusion weapons and at the same time to refine fission weapons for possible battlefield use. One process served the other. As Bradbury explained:

> The first nuclear weapons, the first fission bombs, at least in a limited way, provided the temperatures and pressures which might be adequate for inducing a fusion reaction. Clearly, if we could think of that, so could the Russians. The difficulty was that this was an extremely difficult problem. It was one of the major reasons for making smaller, more efficient weapons; weapons of different internal designs, if I may say so.[27]

At its meeting of July 12, 1950, once again referring to possible "public relations problems," the AEC commissioners authorized a search for a site throughout the North American continent. By this time Lilienthal had left the commission, and the new chairman was Gordon E. Dean, a former law professor at the University of Southern California. The new survey, Dean said, was to be conducted with "a high degree of security" so as "to prevent undue speculation, opposition or apprehension in the area to be examined."[28]

During the summer of 1950, the 5,470-square-mile Las Vegas

Bombing and Gunnery Range northwest of Las Vegas emerged as the favored site. It was a typical slice of the southern portion of the Great Basin, where the first human habitation occurred some 24,000 years ago. The Paiute Indians roamed across the range, leaving behind them the first traces of weapons—carefully honed obsidian arrowheads. Then, starting in the middle of the last century, white prospectors, miners, and ranchers came into the country carrying more efficient weapons, and the Indians were decimated.[29]

Los Alamos was actively involved in the site selection process. Bradbury said:

> Although I find it difficult to cite chapter and verse, our staff looked at maps along with everybody else; and one of the first places we saw was the Air Force base in Nevada, all of which was empty desert. The population problem was almost zero. It was already government property. It was close to sources of technical supply. It was seventy-five miles from Las Vegas. Accessible by air. Accessible by road. And it seemed to us— I would say it still seems—that one could have more effective diagnostic tests there than any place you could imagine elsewhere, and do it safely.[30]

A high-level AEC staff meeting on July 25 reached the conclusion that the Air Force range was the best site. "All agreed," wrote one participant, "that radiological hazards and the public relations problem related thereto constitute the big problem on a continental site and specifically Tonopah." Other sites considered and dropped at the time were the North Carolina and Texas coasts, the Trinity site, Dugway Proving Grounds in western Utah, the Aleutian Islands, Churchill on Hudson Bay in Canada, the Australian desert, and once again the Galapagos Islands—but this time their land mass, where Charles Darwin studied the unusual fauna and flora that helped him reach his conclusions for *The Origin of Species*.[31]

The most extensive discussion concerning the radiological hazards of the Nevada site took place August 1 at Los Alamos. On hand for the all-day meeting in the conference room of the Theoretical Division were those scientists who dealt with either the theoretical or operational aspects of weapons tests. The twenty-four persons in the conference room included Edward Teller, who along with Stanislaw Ulam was soon to produce the hypothetical breakthrough that led to development of the hydrogen bomb, and Enrico Fermi, a Nobel Prize winner who played a key theoretical role in developing the device tested at Trinity.

There was a great deal of hypothetical discussion at the meeting, best summarized by Fermi, who believed that "our conclusions should stress the extreme uncertainty of the elements we had to go on, and that we did our best with these." Fermi thought that the populace should be warned to stay indoors and take showers, and also to take other precautions. The conclusion the participants eventually reached was that the risk "is not a probability that anyone will be killed, or even hurt . . . but it does contain the probability that people will receive perhaps a little more radiation than medical authorities say is absolutely safe."[32]

One of the participants, in fact the scientist who chaired the meeting, was a case in point. Alvin C. Graves, who headed the Test Division at Los Alamos and who later served as test director for most Nevada shots during the 1950s, was a living example of what could happen to a person who was exposed to a harmful dose of radiation. The visible symptoms of radiation sickness that he exhibited after an accident at the laboratory were closely studied by his colleagues. Nevertheless, when the same symptoms occurred in the downwind population, the AEC either ignored them or blamed them on more prosaic illnesses.

Al Graves was an assistant professor of physics at the University of Texas in 1942 when he was invited to the University of Chicago to help build the world's first atomic pile. The physicist came to Los Alamos in 1943, the year the laboratory opened, and two years later he took charge of J Division, as the test group was known. In 1946 Graves was thirty-seven years old, married, and the father of one child.

On May 21 he was standing three feet from Louis Slotin, a brilliant and daring Canadian scientist who was demonstrating to a small group of physicists the behavior of a nuclear chain reaction in sub-critical masses of fissionable material. With a screwdriver, Slotin held apart two hemispheres of plutonium on a rod. He was modeling the interior mechanism of the Nagasaki bomb in a maneuver called "tickling the dragon's tail." The tail twitched, the tool slipped, and the material suddenly became critical. A dazzling blue glare illuminated the room. Acute radiation, as if from an atomic bomb, bathed the eight men. The reaction lasted only a fraction of a second, as Slotin desperately tore the hemispheres apart with his hands.

There was no doubt in Slotin's mind that he would die, but he did not lose his self-control. At his direction, the seven others in the room, including Graves, who was standing closest to Slotin, resumed their original positions, which Slotin sketched on the blackboard so that the doctors could ascertain the amount of radiation each had received. While waiting for a car to take them to the hospital, Slotin told Graves, "You'll come through all right, but I haven't the faintest chance myself." Slotin received a whole-body dose of about 2,000 roentgens, while his hands received a dose of from 3,000 to 30,000 roentgens. For the first five days in the hospital, Slotin was in good condition, but then he failed rapidly and died in agony on the ninth day.

Graves was partly shielded from the critical mass. He received a whole-body dose of nearly 400 roentgens but was told, and subsequently believed for the rest of his life, as did others, that it was 200 roentgens. Bradbury said that Graves had been told that he had received the lower dose "for psychological reasons." The higher dose gave Graves a 50 percent chance of living, while the lower dose held out a greater chance for survival. It was the practice of the laboratory, said its former director, to tell scientists and technicians that they received less radiation than they actually did.[33]

Graves's symptoms were similar to those of the Japanese bomb survivors and the offsite population. He vomited and had a fever but otherwise was not ill during his two-week stay in the hospital. Three weeks after the accident Graves was extremely tired. The hair on the left side of his head came out by the handfuls. It grew back some

three months later. The scientist was temporarily sterile. He had a
red burn over his body. Later Graves developed radiation cataracts
in both eyes. He went back to work not too long after the accident
and participated in the 1948 tests in the Pacific. Three years after
the accident Graves was working and skiing hard. Five years later he
became the father of a healthy son. The other six men in the room
suffered only minor radiation injuries.[34]

Within a year, two men had died and eight had been injured
as the result of this and another accident at Los Alamos. Graves's
close call was widely known within the laboratory, but it was kept
secret from outsiders. Because the irradiated men could be closely
observed shortly after exposure (which had not been the case after
the bombing of the Japanese cities), the data added much to the
knowledge of the effects of radiation. The February 1952 issue of the
Annals of Internal Medicine carried an exhaustive 231-page account
of the accident. Whether by word of mouth or from the available
literature, there was no excuse for AEC personnel not to have recognized the symptoms of radiation sickness. Those symptoms were
a living presence within their own ranks.

During the years of atmospheric tests, Graves toured the small
communities around the test site to assure the inhabitants that there
was no danger from fallout. He testified at the 1957 fallout hearings
of the Joint Committee on Atomic Energy on the chance of radiation
causing cancer. Graves said, "The danger is not that this will happen
to you. The danger is that it is more likely to happen to you. Maybe
the more likely is not very much more likely, but it is still more
likely."[35]

Graves was a mainstay of the Los Alamos community. It was he
who explained to the younger scientists why their work was necessary for the defense of the country. He played the cello in the
Los Alamos Symphony, was an elder of his church, chairman of the
board of the local bank, and a three-term member of the local school
board. Graves died of a heart attack while vacationing in Colorado in
1965. His friend Norris Bradbury said then: "He was one of the pioneers in the field of nuclear energy, having been involved in every

major development in the field from the first atomic pile in Chicago in 1942 and with nearly every atomic test conducted by this country since then."[36]

Following the Los Alamos meeting, where the radiation hazards of the Nevada site were touched upon, the Los Angeles firm of Holmes & Narver conducted a quick engineering study of the site, but little else was done until early November 1950, when President Truman concurred with the commission's decision to seek a continental site. There was no great hurry, said Truman, who up to that time had not been very enthusiastic about the idea.[37]

Meanwhile, Bradbury kept up the pressure for the Nevada site. He was getting desperate. A booster shot, meaning a small thermonuclear explosion that would set off a much larger one, needed to be perfected, and it did not look like a Pacific site would be available. The military had committed most of its manpower and transportation to the Korean conflict. If Eniwetok was unavailable, said Bradbury, "We see no obvious way to maintain this flexibility other than to be able to retire to a United States site on short notice." He strongly recommended the selection and quick preparation of the Nevada site, where "shots would have to be fired in increasing order of yield, and the termination of the program might be indicated by observation of fallout phenomena other than those predicted."

Bradbury outlined the vulnerability of Pacific testing in this rocket to Washington:

We recognize the good intentions of the Chiefs of Staff, but we have no illusions as to the priority which they have granted and will have to continue to grant to "police actions" in Korea and potentially elsewhere. In plain fact, only the USSR can really determine whether or not we can test in Eniwetok in 1951 or at any other time in the near future.[38]

The leisurely search soon gave way to frantic action. On November 25 the Chinese attacked and badly routed American forces in North Korea. A world war seemed imminent. At a press conference President Truman indicated that the use of atomic weapons was under consideration. Actually, consideration of the use of such weapons had been under active consideration since November 10 by an ad hoc committee of the National Security Council.[39]

This was the national emergency that the AEC had specified was necessary for continental tests. Earlier uncertainties and hesitation about selecting a continental site, and the Nevada site in particular, evaporated overnight. The bombing range was government owned and could be made available for immediate use, and there was an existing air base at Indian Springs. A Los Alamos report pointed out that the "'sector of safety' into which the cloud may move with an assurance of safety" extended some one hundred miles to the east and contained only 3,682 inhabitants. The report excluded the population centers of Las Vegas, Tonopah, and St. George from this sector. Unspecified controls could be imposed upon the civilian population to insure safety. The report stated, "In summary, according to conservative estimates based on present knowledge, the Frenchman Flat area offers no foreseeable radiation hazards for detonations possibly as high as 50 KT [kilotons] and certainly none for a 25 KT detonation."[40] (A kiloton is an explosive force equal to 1,000 tons of TNT. The Hiroshima bomb was 13 kilotons.)

On this matter, the government moved, in a relative sense, about as quickly as an atom is split. On December 12 the AEC approved the Nevada site. A special committee of the National Security Council consisting of the AEC chairman and the secretaries of state and defense did likewise three days later, and Truman followed suit on December 18.[41] The president recalled the following conversation with AEC chairman Dean in his memoirs:

"Gordon," I asked, "if we set up this testing ground, will it really help our weapons program from the standpoint of time?"

Dean assured me that it would.

"Can this be done in such a way that nobody will get hurt?" I asked.

Dean said that every precaution would be taken. I told him to go ahead. I suggested, however, that it might be well to do it without fanfare, and very quietly to advise key officials in the area of the plans we had for the testing area.[42]

December and the following months were an exceedingly dark time for President Truman. "I've had conference after conference on the jittery situation facing the country," wrote Truman in his private diary. Korea, the two Chinas, Japan, Germany, France, England, and India—it did not seem like the president could look anywhere without the world threatening to come apart. Truman confided, "I've worked for peace for five years and six months and it looks like World War III is here. I hope not—but we must meet whatever comes— and we will."[43]

This mood of extreme national anxiety seeped down to those who conducted the first tests in Nevada the next month and remained lodged within the weapons bureaucracy for the next few years. Only the president and a few others at the highest levels, however, were privy to information that seemingly threatened the core of this nation's security and put into motion the machinery for a nuclear attack on Russia and China. Though the attack was never carried out, the hasty preparations were indicative of the uncertain times.[44]

In the spring of 1951, with Truman's approval, nine atomic bombs were transferred from AEC control to the military for possible use on Communist troops and equipment believed to be massed on the Asian coast for an attack on Japan. There was no indication that Truman authorized their use, just their transfer to place them in a position of readiness. The incident is described in a recently released section of Dean's office diary, portions of which were deleted by DOE censors.

In late March the civilian chairman of the AEC's Military Liaison Committee, Robert LeBaron, who had regular access to the Joint Chiefs of Staff, told Dean that "the situation in Korea looked rather unfavorable; that the Chinese were massing large forces; that they apparently intended to use their air forces for the first time; that this was coming about with definite support from the Russians; that Molotov had been delegated by Stalin to handle matters in Asia;

that it might even result in a decision by the Russians and Chinese to launch an offensive against Japan."[45]

Things then began to move quite fast. On April 5 Dean learned that the Joint Chiefs of Staff were about to request the transfer from the AEC stockpile to their control of a number of complete atomic bombs. Dean, who distrusted the military, believed that such a request should go through the subcommittee of the National Security Council and that it be made clear that "a decision to transfer was not a decision to use." The AEC chairman attempted to head off a decision to use atomic weapons that would be based solely on military considerations.[46]

Dean was summoned to the White House on April 6. It seemed like a nuclear World War III was imminent. The president told him that "the situation in the Far East is extremely serious; that there is a heavy concentration of men just above the Yalu River in the part of Manchuria across from the northwestern corner of Korea; that there is a very heavy concentration of air forces on several fields and the planes are tip-to-tip and extremely vulnerable; that there is a concentration of some seventy Russian submarines at Vladivostok and a heavy concentration on Southern Sakhalin [Island]—all of which indicates that not only are the Reds and Russians ready to push us out of Korea, but may attempt to take the Japanese Islands and with the submarines cut our supply lines to Japan and Korea." Truman said that the Joint Chiefs of Staff had requested the transfer of the bombs, but "no decision had been made to use these weapons and he hoped very much that there would be no necessity for using them." The special committee of the NSC would be consulted prior to use, Truman said. Dean cautioned that the bombs should not be dropped in North Korea, where, because of the terrain, they would be "ineffective and psychological duds." Truman agreed.[47]

But the president was greatly concerned about keeping secret what he had authorized. Three days later he called Dean and said he knew that it was the AEC's obligation to report such a transfer to the Joint Committee on Atomic Energy, "but he thought it highly important that word of this transfer not get to the full committee at this time; that this was a highly sensitive military move designed to secure readiness and it was not something that should be thrown into debate in Congress." The president suggested that he would inform the committee chairman, Senator Brien McMahon, so that the AEC would not be caught in the middle. Dean assented.

At a meeting of the joint committee on April 9, Dean was asked if any nuclear components or assembled atomic weapons were outside the United States. He answered no to both questions but felt bad about deceiving Congress. The nine bombs, all of them of an early model known as the Mark 4, were transferred that same day to waiting Air Force planes and presumably flown across the Pacific.[48]

A few months later, hearing that the president was under the mistaken impression that artillery capable of firing atomic shells was ready for use in Korea, Dean asked for a meeting with him. The AEC chairman explained that no such weapon was then available, and the earliest it could be developed was the fall of 1952. Dean stated in his diary, "I emphasized, however, our capacity for the existing [nuclear] weapons to employ these effectively at any moment against troop concentrations." Truman then told Dean that he would like to witness a test at the Nevada site, but he never made the trip.[49]

Exasperated at the inability to negotiate a Korean truce in early 1952, Truman let off steam in his diary against the duplicity of Communist governments. Truman wrote, "It seems to me that the proper approach now would be an ultimatum with a ten-day expiration limit informing Moscow that we intend to blockade the China coast from the Korean border to Indo-China, and that we intend to destroy every military base in Manchuria, including submarine bases, by means now in our control; and if there is further interference we shall eliminate any ports or cities necessary to accomplish our peaceful purposes." Truman cited a long list of past Russian transgressions, ranging from the kidnapping of children in occupied countries to the supplying of war materials "to the thugs who are attacking the free world." Then he outlined this drastic remedy:

> This means all out war. It means that Moscow, St. Petersburg, Mukden, Vladivostock, Peking, Shanghai, Port Arthur, Dairen, Odessa, Stalingrad and every manufacturing plant in China and the Soviet Union will be eliminated.

This is the final chance for the Soviet Government to decide whether it desires to survive or not.[50]

When he was out of office and in a calmer mood, Truman said he had been pressured to use nuclear weapons by persons who did not understand that they were of little value except on relatively large centers of population. The former president added, "It could have been used to destroy Peking or Vladivostok or any of the Manchurian cities, but we were trying to avoid that necessity, and we did avoid it."[51]

It was within this highly charged atmosphere that preparations went forward for the first tests. On the day following the president's approval of the site, a "public relations conference" was held. Attending were a general from the Military Liaison Committee, the State Department's expert on atomic matters, and a member of the AEC's Public and Technical Information Service. Press releases, it was determined, should stress that "It had been done before and we can do it again." Item four of the eleven items under discussion was: "It appeared that the idea of making the public feel at home with neutrons trotting around is the most important angle to get across."[52]

From December 19 to January 11, 1951, during one of this country's gravest crises, the wording of a one-page press release announcing selection of the site occupied the time of top military and civilian decision makers. A number of drafts were considered. The Joint Chiefs of Staff wanted to eliminate any reference to "radioactive dangers," and the AEC field office in New Mexico, which watched over the Los Alamos operation, warned that too much discussion about the lack of danger from radioactive fallout could come under the category of "the lady doth protest too much." The Military Liaison Committee wanted "a small and somewhat misleading announcement; no reference to intensive tests, eliminate names and radiological on page 2."[53]

The National Security Council eventually became involved in the wording of the press release, as did White House press secretary Joseph Short and finally President Truman, who eventually approved it. The joint committee was notified, the mayor of Los Angeles was called, and the governors of Nevada and the adjacent states were either telephoned or visited personally by AEC officials. Numerous assurances of safety were given to one and all.[54]

The press release was issued at 3 P.M. on January 11, one day after the president approved Operation Ranger, the first test series at the site. It contained the statement: "It has been found that the tests may be conducted with adequate assurance of safety under conditions prevailing at the bombing reservation." A Las Vegas newspaper assured its readers that there was no danger and that they would neither see nor hear the explosions when they occurred. It was wrong on all counts.[55]

CHAPTER 6

THE TESTS

SIXTEEN DAYS after the announcement of the site selection, a one-kiloton bomb was dropped from an airplane and detonated over Frenchman Flat. A portion of the cloud passed over Groom Mine. Radioactive dust and more clouds passed over the mine on subsequent shots, and the Sheahan house was shaken and the door blown open. The first tests were conducted in "a jury rig fashion," admitted Dean, who instituted a press policy of not commenting unless questioned and then leaving it "completely fuzzed up."[1]

Bradbury invited his old physics professor, Ernest Lawrence, to witness the third drop, "which should be a nice sight," he wrote. He gave Lawrence directions about where to meet him and recommended that he wear warm clothes. If the test was delayed, said Bradbury, there were always "the dubious delights of Las Vegas." The laboratory director, who had thought about four-leaf clovers and rabbits' feet before the Trinity test, said: "The only requirement that we make of witnesses is that they join us in keeping their fingers crossed or in making use of any other good luck charm which they feel is reliable!"[2]

Within a few days of the end of the first test series in early February, the first claims were submitted for broken windows. Senator McMahon, a former law partner of Dean's, called the AEC chairman and joked that the agency would have a claim for every window broken in Nevada during the last twenty-five years. Dean thought that for the sake of public relations the claims should be paid quickly from a congressional appropriation or a secret AEC fund, if there was one that could be tapped for this purpose.[3]

Operation Buster-Jangle, which followed in the fall and early winter of 1951, included three surface shots for the first time. Surface shots, which sucked up dust and other debris, were generally dirtier than air bursts. In November, researchers from the University of California at Los Angeles, who were living in a stone cabin north of the test site while they collected soil samples, were doused with fallout. The project leader, Kermit H. Larson, recalled: "On that particular day I accumulated, myself, a little over 5 r and that was enough, we thought, and we went back home and talked about our bloody wounds and so forth and came back out for Jangle and at that time we had a good contamination." The Sheahans and others who lived nearby did not have the luxury of knowing what dose they had accumulated, or what effect it might have.[4]

A Los Alamos Test Division memo noted, "The Jangle test program has raised for the first time since Trinity serious problems of radiation safety at moderate distances from the test site." There was a great deal of technical discussion, but no conclusions were contained in this report and another one relating to the series.[5]

The tempo of the nuclear weapons program increased markedly in 1952. In mid-January, President Truman called a meeting of the chairmen of the AEC and the Joint Chiefs, the secretaries of defense and state, the head of the Bureau of the Budget, the Director of Defense Mobilization, and the secretary of the NSC to discuss a greatly expanded nuclear weapons program. The international situation remained tense, and there were not enough bombs to cover all the targets in the USSR. Truman went around the room to get the opinion of each person. He then gave a general statement in favor of the expansion and asked if anyone in the room objected. No hands were raised, so the program, which would mean greatly increased testing in the next few years, went forward.[6]

Just before the next series of tests got underway, Chairman Dean complained to a House subcommittee that "we are once again 'under the gun.'" Construction was still going on at the half-completed site, and the military was adding additional tests to the series. The atmosphere at the test site was harried. Plumbers were working more than 120 hours a week, and workmen were sleeping eight to a room. The kitchen facilities were far from ideal; there were forty cases of dysentery in one week.[7]

Operation Tumbler-Snapper got underway with a series of air drops in April 1952. At a press briefing that month, Lieutenant Colonel James B. Hartgering, the radiation safety officer, said, "The residents of this area are becoming old hands at this problem. Through their most helpful cooperation we have been able to insure complete safety for everyone."[8] But May and June were different. They proved to be precursors and distinct warnings of what was to come the following year in terms of exposure of humans and animals to radioactive fallout.

The cloud from Shot Easy on May 7 moved north and passed over Tempiute, a small mining community adjacent to Lincoln Mine and just north of the test site. Some 110 men, women, and children went about their early-morning business as usual as the dust cloud passed overhead. Although radiation monitors were present, there were no warnings or precautions. The fallout over the hillside community was the greatest yet recorded for an inhabited area.[9]

The Sheahans, who had been urged by AEC monitors to vacate nearby Groom Mine when Shot Easy was detonated, could not return via the normal route because there was too much radioactivity. They were told by an AEC official, "You couldn't go in there and stay without getting radiation sickness."[10]

The cloud drifted north and passed over the town of Ely, Nevada. Curving in a gentle arc to the northeast, it then passed over Salt Lake City, where a local manufacturer of Geiger counters noticed unusual activity on his instruments and contacted the governor, who wrote an angry letter to the AEC chairman. Dean referred the letter to Gordon Dunning in the Division of Biology and Medicine for an answer. There were worries that such criticism might limit use of the test site. In fact, Dr. George A. Spendlove, the Utah State Health Officer, had already cooperated with AEC officials in helping to suppress an unfavorable news story.[11]

Some cattle, grazing about twenty-five miles from ground zero, were burned on their backs by the fallout in early June. The symptoms were the same as those exhibited by the Trinity cattle, and AEC officials privately agreed that the cause was fallout. One of the veterinarians who investigated for the AEC was R. E. Thompsett, who was familiar with the Trinity cattle. Gordon Dunning subsequently wrote a report on the incident. Both would be heavily involved in a much more serious livestock incident the following year. The rancher, Floyd Lamb, demanded compensation, but none was forthcoming.[12]

The 1952 fallout episodes were serious enough to warrant a short study on the suitability of the Nevada Test Site. An AEC committee, composed mainly of persons who were responsible for conducting the tests, met in January 1953 to consider the future of the test site. Preparations were well underway at this time for the Upshot-Knothole series. It was not likely that there would be any major changes in location or procedures at this late date.

The committee received a letter from a worried Bradbury—ever the protector of his laboratory's interests—who thought that the medical and public relations aspects of the review were being overemphasized. Bradbury then appeared before the committee and stated that there was no more favorable site on the continent. "We must either live with no continental test site, or live with this one under what circumstances are permitted. Needless to say, I hope that a firm stand as to the technical necessity of a continental site can be maintained in spite of the understandable nervousness of those whose major objective is amiable public relations," said Bradbury.

The man in charge of public relations for the test site, Richard G. Elliott, told the committee that his goal was to secure "the understanding and cooperation of the site region public so that unthinking public reaction could be removed as a pre-shot consideration." The offsite public, Elliott said, have "rather gingerly accepted scientists' assurances of non-hazard as to fallout." He added that radiation continued to pose a threat to effective public relations.

A suggestion by the military to hold the tests under the worst possible conditions, such as "very heavy rain" or snow, was quickly dismissed by the committee, which noted: "There was considerable discussion of the improbability of obtaining heavy weather at either Eniwetok or Nevada within a schedule and to the considerable hazards which would result if such conditions were used in Nevada."

The committee's decision to continue use of the test site was based mainly on Bradbury's argument that there was no alternative. The committee congratulated itself on "the good fortune which has on occasion caused highly radioactive clouds to wend their way in between communities." That good fortune was to end with the Upshot-Knothole series.[13]

Upshot-Knothole was first conceived in late 1951 as a test series of weapons effects with not more than three shots. It had grown to a ten-shot series when routinely approved in February 1953 by President Dwight D. Eisenhower, who had just taken office. The ten scheduled shots were crowded into a short time, and an eleventh shot was added at the last moment. Seven of the shots were the dirtier surface variety, which were detonated from towers. Los Alamos was responsible for all of the tests except two, which were duds.

Those two tests represented the entry of the Livermore Laboratory into competition with its more established rival for time, manpower, and materials at the test site. Of the duds, Livermore director Herbert York later wrote, "Some Los Alamos scientists filled the air with horse laughs." There would be other shots that fizzled over the years. Of the two laboratories, generally Livermore was more venturesome with its experiments.[14]

There was a sense of continued urgency in the spring of 1953. The Chinese were thought to be poised for another attack in Korea, the Russians were being aggressive in Berlin and the Arab world, and the United States believed that it was in a race with Russia to produce a deliverable hydrogen bomb. Most of the shots in the series that was to produce the most fallout offsite were aimed at perfecting the fission device that would detonate the fusion bomb to be tested the following year in the Pacific.[15]

There was also a lighter side to the series that masked its more deadly intent. A sense of entertainment, fostered by the media, pervaded Upshot-Knothole. The press was allowed to cover the first shot, and six hundred observers were on hand. Chet Huntley crouched in the trenches with the troops, and Walter Cronkite, back at News Nob, helped portray the test and the fate of mannikins seated in two typical homes, dubbed "Doom Town," to some 15 million television viewers. The mannikins, amidst reconstructed surroundings, and automobiles used in the test were displayed later that year in downtown Los Angeles.

Nevada Highways and Parks magazine had an extensive layout on the observation shot. The magazine, whose purpose was to promote the state, told its readers that the AEC said there was no danger from fallout on the state's highways, and then added this not untypical description of a shot:

> All of the blasts are frightfully terrible yet unbelievably magnificent; they are hellish but beautiful; horrible yet spectacular. The whole range of human emotion is brought into play upon observing such a detonation. From a distance each appears as a gorgeous fireworks display on a gigantic scale. Nagasaki and Hiroshima tell the story differently, however, on what could happen should one be dropped on an American city by a relentless enemy.[16]

There were these unusual activities. At the second test, nine military officers withstood the blast only 2,500 yards from ground zero. Two of the officers said they would return, if given the chance, and witness a test from 2,000 yards. A featured dancer at one of the Las Vegas hotel-casinos performed an "atomic ballet" that was recorded by a newspaper photographer atop Mount Charleston, from where the tests were visible. Back in town there was an Atomic Hairdo, an Atomic Cocktail (equal parts vodka, brandy, and champagne, with a dash of sherry—then, bang!), and the Atomic Bomb Bounce, a popular dance.

A 280-millimeter cannon fired an atomic projectile on May 25, and the drumbeat of publicity that accompanied its journey across the country and arrival at the test site obscured the grave problems with Shot Harry. It was as if the public and the media, tiring of the theme of awesomeness, turned to the freakish aspects of the tests, and the AEC, for its own purposes, went along with the frivolity.[17]

The only advance warning that there might be fallout that spring was given on a restricted basis to the National Association of Photographic Manufacturers, whose secretary was authorized to give the association's membership the approximate starting date, number of tests, and expected length of the series. As far back as the Trinity test there had been problems with radiation-drenched raw materials exposing film and light-sensitive printing paper. In 1952 members of a technical committee were given warnings, and in 1953 that advantage was widened to all members of the trade association. No such warnings were given to the offsite public.[18]

But there was an extensive public information program for the test series approved by the commission in February. One of its purposes was "to allay unfounded fear of damage or injury that may arise from public misunderstanding of test operations." To explain those operations, the AEC put out a pamphlet entitled *Continental Weapons Tests . . . Public Safety*. The booklet stated that "there has been no instance of harmful exposure of human beings to radiation from fallout, and only a single reported instance of observable radiation effects on cattle (grazing immediately adjacent to a firing area)." There was no mention in the booklet of the possibility of cancer developing in future years, although the AEC knew at this time of the growing number of leukemia cases from the Japanese experience. For the three series to date, it noted, 453 claims had been filed for structural damage—such as broken windows and cracked plaster—and settlements totaling $43,000 were reached in 89 percent of the claims.[19]

The first serious fallout episode that spring followed Shot Nancy, the second in the series. Fallout again descended on Tempiute during the early morning hours of March 24. The UCLA research group, which was working out of one of the cabins at Groom Mine, was warned by radio the previous evening that the fallout was aimed in their direction. They placed the rabbits that were being used in an experiment in position and departed "because they did not believe in taking any excess amount of radiation," said mine owner Sheahan.[20]

At Tempiute the AEC asked residents to remain indoors, but

the warning was not given in time. Those who were outside felt a pressure in the air. There was a metallic taste, their eyes stung, and exposed skin was burned. The AEC investigated and took reports of symptoms of radiation sickness from five persons in the community of 175. The maximum dose was 3.4 roentgens, the AEC estimated. Between Tempiute and Ely thousands of sheep were grazing on their winter range. The cloud again passed over Ely and headed toward Salt Lake City. None of the radiation levels were hazardous, said the AEC.[21]

After Shot Nancy, the AEC hurried another pamphlet into print. Designed as a supplement to the one issued earlier in the spring and to answer the increasing number of questions from the public, the booklet stated, "During the present and past series no community has been exposed to hazardous radiation." However, the pamphlet noted that rats, mice, rabbits, dogs, monkeys, pigs, and sheep were being used in experiments. (In time, some offsite residents came to believe that they were being used as guinea pigs.)[22]

Twenty-four hundred soldiers, fourteen members of Congress, eleven generals and admirals, three captains of industry, two officials from the Department of State, one university president, the mayor of Los Angeles, and one AEC commissioner were on hand on April 25 when Shot Simon exploded with unexpected vigor. The scientists had underestimated the yield of the device. A drone plane flying too close to the blast disappeared "in a flash and a puff." Rocks whizzed past troops crouched in trenches 4,000 yards from ground zero, and radioactive dust poured down their necks. Sagebrush three miles away burst into flames. The VIPs watched in stunned silence as the shock wave, resembling the undulation caused by a rock thrown into a still pool, moved rapidly toward them. They were knocked backwards, their hats were blown off, their eardrums pierced by a loud riflelike crack, and all the windows in their busses were blown out. They were checked with Geiger counters and were moved out of the area before the radioactive dust cloud enveloped them. One congressman remarked, "It was an unimaginable experience!" A harried Alvin Graves, the test director, gave them a quick briefing and then went back to trying to contain the offsite damage.

The fallout cloud from Simon drifted southeast, and fallout descended on a sparsely populated region around Mesquite, Bunkerville, and Riverside, Nevada. When monitors on state highways measured heavy amounts of radioactive fallout, vehicles were halted at roadblocks, checked for radiation, and washed for the first time in the short history of testing. The AEC publicly reported as late as twelve hours after the detonation of Simon "that no fallout had as yet been reported from any inhabited localities." But the term *inhabited localities* was a tricky designation. Within the local fallout area, where a maximum of 15 roentgens was measured, some 1,400 persons lived in scattered locations. The high reading was taken at the Riverside Motel, just off the main Las Vegas–Salt Lake City highway. The motel was inhabited by about a dozen people. The AEC said there was no danger.[23]

The fallout cloud from Shot Simon, however, was headed toward more densely populated regions. The cloud proceeded east at a steady pace, and the AEC later noted: "Fairly large amounts of radioactive dust fell out over about half the United States after burst 7." During a rainstorm thirty-six hours after detonation, fallout settled onto upstate New York in a band from an undetermined point north of Troy to Poughkeepsie. The alarm was given when radiation counters went off in the nuclear research laboratory at Rensselaer Polytechnic Institute. An estimated area of 7,000 square miles, centering on Albany, received radiation from the rainout.[24]

On May 13 the commissioners were given two pieces of bad news. First, they were informed that representatives of Great Britain and Canada, who had met with the AEC staff six weeks previously, had refused to endorse the AEC radiation safety requirements for onsite workers and the offsite population at the Nevada site. The guideline was 3.9 roentgens exposure over a thirteen-week period. The same amount of radiation could be acquired within one week, provided there was no further exposure. The Commonwealth delegates suggested a more restrictive guideline of .9 r per week. John C. Bugher, who headed the AEC's Division of Biology and Medicine, said the lower level "would make it impossible to conduct operations at the test site without major changes in present procedures." Bugher voiced confidence in the higher number, and the standard was not changed. The second piece of bad news concerned Shot Simon. The commissioners were briefed on the fallout over New York. The reason for such a large amount of fallout from Shot Simon, AEC commissioners were told, was that the detonation had "considerably exceeded" its estimated yield. A maximum dose of 2 roentgens was measured in the East—a fair amount, considering the distance the fallout had traveled, but below the 3.9 r guideline, the staff noted.[25]

The fallout did pose a hazard, however. The danger to children from ingestion of radioiodine in milk as a result of Shot Simon was later estimated in an article in *Science* magazine by Ralph E. Lapp, a nuclear physicist and an early critic of the AEC. Lapp calculated that the thyroids of 10,000 infants in the Albany-Schenectady-Troy area could have received from 10 to 30 rads, enough to produce from 10 to 100 cases of thyroid cancer over the next twenty years.

Lapp was no stranger to the ways of the AEC. The consultant and writer, who had worked in the Navy's nuclear program in the late 1940s, published an article in the *Bulletin of the Atomic Scientists* in 1955 that discussed the radioactive effects of the hydrogen bomb. The article was based on a careful reading of the available public literature. At the request of Senator Bourke B. Hickenlooper of Iowa, who was disturbed by "the startling accuracy" of the story, the AEC launched an investigation of Lapp, suspecting that he had used classified material. The investigator who interviewed Lapp was convinced that he had used unclassified materials. Two years later, as one of the few AEC critics allowed to testify at the joint committee hearings on fallout, Lapp faced the hostile questions of its members.

Lapp's early criticisms of the AEC were predated by those of Lyle Borst, who threw a scare into the AEC bureaucracy during the Upshot-Knothole series. Borst, who had formerly worked in the AEC's Brookhaven National Laboratory in New York and who was now at the University of Utah in Salt Lake City, wrote a letter to Gordon Dean after Shot Simon, stating that he was greatly concerned over "the long range effects of AEC policy." Borst added, "When I find contamination on my children the equivalent of any contamination I have ever received in eighteen years of nuclear work, I cannot consider it inconsequential nor trivial." He said he would keep his children indoors when his monitoring equipment indicated there was fallout in the air.

Lapp pointed out the discrepancy between safety measures in the AEC laboratories and the lack of precautions for the offsite population. "In my AEC experience we have always removed pregnant women from contact with radiation—from lower levels than are now assured to be safe. Yet women who happen to live near the test site are given no such consideration," wrote Borst, who made his concerns public through the newspapers. The commission considered the matter, and Dr. Bugher sent a letter in reply, which ended: "We find no basis for concluding that harm to any individual has resulted from radioactive fallout."[26]

By early May the pressures of the crowded, problem-plagued test series began to tell on those who were conducting the tests. Replying to internal criticism, a frustrated Carroll Tyler, the test manager, told his superior in Washington:

> In general, it must be recognized that it is impracticable to conduct, with sound economy of effort and money and with maximum utilization of the proving ground toward expeditious forwarding of weapons development, a series of tests, conditioning each shot on weather, atmospheric and technical characteristics which insure no possible radiation or blast hazard to any persons or communities. It is recognized and practiced that these are key factors in deciding whether or not we will fire, but they are and must be weighed with many other considerations and particularly against the possibility of a really accurate forecast and with acceptance of the fact that we are not fully informed, though we are gaining knowledge rapidly, as to what in detail the effects will be under any particular set of conditions.

Tyler, who was the AEC employee responsible for deciding whether to fire or postpone a test, pleaded for a relaxation of the radiation safety guideline, which had already been considerably exceeded in some locations around the test site that spring. He said, "This does impose a serious and perhaps unrealistic limitation upon an expensive, important, and onerous operation."[27] Tyler's classified message differed markedly from what the public was being told by the AEC. The gulf between public assurances and private doubts was to widen considerably after Shot Harry.

Following Shot Harry, the fallout on a public and private level was both loud and profuse. The hazards of Harry were downplayed, however, and Los Alamos scientists pushed hard to add an unscheduled eleventh shot to an already crowded series—an airburst that would be twice the yield of Harry. The single test would obviate the need for a fall series and was crucial to the successful testing of a hydrogen bomb in 1954. It was the triggering device for that weapon.[28]

Graves attempted to assure the commission. He said that since this was the last shot in the series, "it would be possible to select weather conditions with great care." The implication was that weather conditions for the prior tests had been chosen with less care. Two of the commissioners pointed out that the absence of any serious fallout on populated areas "although due largely to careful planning of the shots, was also due in some part to good fortune."[29]

Commissioner Eugene M. Zuckert, citing the St. George incident and mounting claims for sheep deaths, said he was concerned about the eleventh shot. His concern was not because of any possible effects on human and animal health but because it might set the program back if there were problems. He pointed out that the letter of request to President Eisenhower "did not inform him of the magnitude of the shot or the possible dangers involved." Zuckert thought that Lewis L. Strauss, a former AEC commissioner who was Eisenhower's special assistant on atomic energy matters, should be informed of these dangers.[30]

Chairman Dean wrote and talked to Strauss four days later. The gist of his remarks was that the commission felt that this shot was essential to the thermonuclear program. Strauss, who sympathized with the need for the test, assured Dean that he would get the matter to the president, who approved the test the next day.

Dean's diary for May 27 noted:

The President expressed some concern, not too serious, but made the suggestion that we leave "thermonuclear" out of press releases and speeches. Also, "fusion" and "hydrogen." If something comes up and we want to use it, we can get clearance from Cutler [Robert Cutler, head of the National Security Council]. The President says "keep them confused as to fission and fusion."[31]

Dean passed these instructions on deception to Morse Salisbury, director of public information for the AEC, and they were implemented.[32]

There were no unusual problems with Shot Climax, which weighed in at 61 kilotons. On June 4, Strauss slipped a handwritten note on White House stationery to Eisenhower, stating, "The concluding atomic test of the current series was successfully staged this morning." A short time thereafter Strauss took over as chairman of the AEC with instructions from President Eisenhower to seek nuclear disarmament. Strauss, a self-made Wall Street millionaire and former Navy admiral who had strong opinions and a strong dislike of communism, did not heed this presidential directive during his five-year tenure as AEC chairman.[33]

Executive control over atomic affairs was tenuous. On matters of what weapons were tested and how they were tested, the role of the two presidents who were in office during the years of atmospheric activity was negligible. (The same could be said of those who followed, as we shall see.) Truman and Eisenhower were both protected from some of the harsher realities of national defense, and both routinely approved test shots with hardly any knowledge about or concern for local fallout. Both considered visiting the test site, but neither made the trip. Politically, it was best for a president to appear close, but not too close, to the means of mass destruction and the lingering aftermath.

Under Truman, the nation's nuclear arsenal rose from almost no weapons to 1,600 by the time Eisenhower took office in 1953. Eisenhower was horrified by the amount of destructive power available to him, but the cornerstone of his military policy was fewer troops and more such weapons. This meant continued testing. In response to a call in the spring of 1956 for an end to testing from Adlai Stevenson, a candidate for the Democratic presidential nomination, Eisenhower said, "Research without tests is perfectly useless, a waste of money." In his second term Eisenhower became worried about the health effects of global fallout. This concern led to his declaring a moratorium on tests in 1958 over the vehement protests of Strauss, who, in company with Ernest Lawrence and Edward Teller, advised the president on technical and policy matters relating to nuclear weapons. It was not until later in his second term that Eisenhower placed independent science advisors in his own office. The subjects of weapons testing and radioactive fallout were too technically complex and arcane for any president to master, except at the general policy and political levels. Thus, by abdication, bureaucrats ruled upon these matters.[34]

Following the Upshot-Knothole series, the AEC commission dealt with the problems of fallout by embarking upon a stepped-up public relations program after being told that the offsite residents had little faith in what the AEC said. Once again, public relations was to be the cure-all.

At its June 17 meeting the commission was told that a resolution had been introduced in the Senate calling for a halt to further continental testing. The resolution was referred to the Joint Committee on Atomic Energy, where it died a silent death. At the time, the chairman of the committee, Representative Sterling Cole, was writing soothing letters to worried citizens, saying that the fallout posed no danger to health and was not modifying the weather.[35]

Acting at the suggestion of Morse Salisbury, the commission approved the production of a film about the St. George fallout episode, subject to its approval of the script. Salisbury said he had mentioned the film idea to Congressman Stringfellow, who liked it and suggested that the premier be held in St. George. Salisbury later noted: "The film was designed as part of the education program to dispel the unwarranted worry among residents in Nevada and adjoining states about hazards from tests. This worry was threatening continued use of the test site."[36]

The movie, an idealized version of the events on the day of Shot Harry, was made that fall. Frank Butrico was brought back to St. George for a few days to portray himself. The narration implied that all the citizens of St. George took cover in time to escape the fallout from Harry. The narrator intoned, "Actually, when the invisible cloud had passed, the total amount of radiation deposited on St. George was far from hazardous. Then, you may ask, why were people asked to stay indoors? For a very simple reason. The Atomic Energy Commission doesn't take chances on safety." The film repeated the basic AEC message: There was no danger; there was no hazard.[37]

The movie returned to haunt the government a quarter century later. The Department of Energy held a workshop in 1980 for offsite monitors in order to gain information that could be used to refute the allegations of the Allen suit. The film was shown to jog the memories of the former monitors, whom DOE officials urged to be candid. Harry Jordan, a monitor in a small Nevada town during the 1953 series who still worked at Los Alamos, prefaced his remarks by noting that they might sound like heresy. He then said:

> The objective of the so-called offsite [monitoring] projects was to establish the fact that the assumption originally made, that there was little or no risk, was indeed true. I think that most of our work, in essence, shows that. I know, however, and I detect that when we show that film and the blunt statement is made that no harm resulted to these people as a result of fallout, that even within this room there are people that do not believe or do not accept that as a true statement today.

Another PHS monitor, Oliver R. Placak, added later at the workshop: "You can't underestimate the importance of public relations when you are trying to dump radioactive material on people [the transcript noted laughter at this point], and we worked at it strenuously."[38]

The idealized film made the AEC and its personnel appear to have performed much more effectively than they actually did on the day of Shot Harry. Such sleights-of-hand were a way to reassure the public and cover tracks. National security and personal careers were protected at the same time.

The film was shown widely throughout the downwind region. The AEC figures may have been inflated, but here they are: 500 persons saw it at the theater at the Lincoln Mine, 400 at St. George High School, 230 at the elementary school, 200 at the LDS Ladies Relief Society, 180 at Dixie Junior College, 160 at Kanab High School, 45 at the Kanab PTA, 35 at the Orderville PTA, 12 at the St. George Fire Department, and so on for a total of 7,500. (In the margin of the AEC memorandum describing this activity, opposite the statement that "practically every person in offsite area saw the film or heard a monitor speak," was the handwritten notation: "I doubt this.")[39]

AEC officials fanned out across the region and gave numerous talks designed to reassure the public. Graves, the man who had survived radiation, was the star attraction. He had the scientific credentials to sound convincing, and he was. To counter Borst's criticisms, Graves met that summer with 300 civic leaders and opinion leaders in Salt Lake City and addressed a joint session of the Utah legislature. "This single experience was most effective in answering the concern and questions of opinion leaders," noted Richard G. Elliott, who was in charge of public information for the test site. Graves also met with University of Utah faculty members and doctors in the Rocky Mountain region. "The medical people appeared 'quite relaxed' about radioactive fallout," said Elliott.[40]

With public relations being emphasized more than ever, Elliott's activities were becoming quite noticeable. Public relations was a new but rapidly expanding concept in midcentury, and Elliott had grown with it. He had entered the profession via the traditional route. Elliott attended a year of college in Kansas before getting a job as a newspaper reporter in Topeka. He then held a series of jobs on small papers in Arkansas, Texas, and Oregon. In Chicago he got a public relations job after lasting one month on the *Herald American*. During World War II, Elliott served in a public relations capacity in the Pentagon. After the war, he opened up his own public relations business in Chicago but did not do well. So when he was offered a job with the AEC, Elliott took it.[41]

Because of the extreme secrecy of the program, the public information director's role was of great importance. There were few other conduits for information. Elliott, who held that job throughout the years of atmospheric testing, was liked and respected by such veteran shot observers as Gladwin Hill, the Los Angeles bureau chief of the *New York Times*. When a pending test was announced and newsmen were to be bussed to the site in the early morning hours, Elliott would turn on a signal light that night in his office to indicate that the test had not been canceled because of poor weather. Newsmen appreciated that thoughtfulness.

When there was no preshot announcement and newsmen suspected a test was upcoming, they watched to see if Elliott left the Hotel Cortez in the early morning hours wearing boots. If he did, this meant that he was on his way to the test site for a preshot briefing. Some of the smarter newsmen also learned of an upcoming test from federal aviation authorities who had to divert air traffic out of the area. The reporters then hotfooted it up to the top of Mt. Charleston for a view of Yucca Flat. Through such cat-and-mouse techniques the world learned about the tests.[42]

At the Allen trial, Elliott, retired from the AEC for some years and in his midseventies, had not abandoned his role as a public relations advisor. While waiting to testify, he listened to Bradbury being pressed on the witness stand to admit that he knew the direction of the prevailing wind at the test site. "Duck that," whispered Elliott. But the former director of Los Alamos was too far away to hear.

The official version of the 1953 test series was contained in the AEC's fourteenth semiannual report, which was released to Congress and the press on July 30, 1953. Its main conclusions were that no radiation levels had been high enough to cause a health hazard and that no claims of radiation sickness had been substantiated. For those persons who thought they had suffered such symptoms, the report stated, "These and other persons who showed apprehension regarding test radioactivity were given factual explanations of the observed radiation levels to alleviate their concern." For the entire series, the average dose a person received in St. George, according to the AEC, was 3.5 roentgens—in other words, within the AEC guideline. There was no mention of a maximum dose, although such doses had been measured at scattered locations. The Public Health Service and the military estimated higher average doses for St. George, but those reports were classified.[43]

Then the AEC once again formed an internal committee to reconsider the suitability of the test site. The Committee To Study the Nevada Proving Grounds began its reevaluation of the site in the summer of 1953. Like the briefer effort earlier that year, the committee's composition was a clear indication of the conclusion it would eventually reach. Bradbury, Graves, Tyler, Bugher, Elliott, and Salisbury, along with four other persons who were also intimately involved with the test program and its implementation, reevaluated their own performance. Gordon Dunning was listed as an advisor to the committee.[44]

The value of the various working papers that comprised the report, which was classified as secret, lies in what they reveal about internal AEC thinking at a time when Operation Candor was being launched. The idea for Operation Candor—first broached by Oppenheimer, then endorsed by Eisenhower, and quickly shot down by Strauss—was to slow down the arms race by telling the American people the truth about atomic affairs.[45]

What the American people were not told, among other things, was what Howard L. Andrews of the National Institutes of Health, an authority on radioactive fallout, wrote in a supplement to the site reevaluation report entitled "Residual Radioactivity Associated With the Testing of Nuclear Devices Within the Continental Limits of the United States." Andrews listed a number of considerations, among them the small safety factor inherent in the AEC standard, the meager scientific data on the subject, the fact that "radiation induced cancer may require 10–15 years to become evident," and the greater sensitivity of children and pregnant women to radiation.

Because of all these factors, warned Andrews, "it seems necessary to adopt a rather conservative attitude toward the involuntary exposure of the general populations. *An error on the radical side will not be immediately apparent, but the chickens will inevitably come home to roost at some later date"* (emphasis added).[46]

By December, to no one's great surprise, Carroll Tyler, who co-ordinated all the reports being written by the individual committee members, told his superiors at AEC headquarters in Washington that the committee would recommend continued use of the Nevada site and "will not impose insurmountable limitations on future use."[47] In addition to being test manager, Tyler was head of the AEC field office that oversaw the operations of the test site and Los Alamos.

The final report, classified secret, was submitted in February 1954. The committee completely discounted the validity of all reports of radiation sickness. The report stated, and underlined for emphasis: "There has been no instance of injury or death of people, and those most intimately aware of radiation's biological effects have concluded that there is no reason to assume that fallout has at any time really endangered people in the sense of immediate or of lingering effects." What Andrews had to say was completely ignored.

The more startling admissions received short shrift. Four weaknesses in the reporting procedures of radiation levels were briefly cited. First, the decision at the start of the series to announce radiation levels for populated areas after each shot was canceled after the first high reading at Tempiute approached the maximum permissible level. Public statements were then confined to reassurances. Second, a plan to tell livestock owners of radiation levels was abandoned, and ranchers were expected to determine the levels themselves and report them to the test organization—not a very workable solution. Third, there was no way to determine the exposure of persons living in isolated areas. Fourth, and most startling of all, was "the delay —perhaps unavoidable—in determining where there is 'significant fallout'" on populated areas—a reference to the "barn door" syndrome. What these admissions amounted to was a disguised confession by those who conducted the tests that what meager actions had been considered for the protection of the offsite public were either abandoned before implementation or were inadequate for the situation at hand. The after-the-fact radiation measurements, such as Butrico made in St. George, were a badly flawed method for protecting the public. But nothing was changed.

The lengthy report, to which twenty-four attachments were appended, was the first thorough, although biased, examination of the safety of the Nevada site. As for the future, since $20 million was already invested in the Nevada site, there was no reason to seek another testing ground. Public relations procedures needed to be strengthened. And, the report noted ominously, the two laboratories wanted "bigger bangs."[48]

The directors of the AEC's Divisions of Military Applications, Biology and Medicine, and Information Services concurred in the committee's findings. On February 17 the AEC commission released funds for further construction at the test site and approved preparations for the next test series, to be conducted at the Nevada site in 1955. Commissioner Thomas E. Murray said "the progress of the program and considerations of economy required the continued use of the Nevada Proving Grounds."[49]

No tests were conducted at the continental site in 1954, that activity having been shifted to the Marshall Islands, where six tests— all but one in the megaton range—were held during the spring. Following the thermonuclear Bravo shot, for which Upshot-Knothole was preparatory, fallout settled upon Marshall Islanders and some Japanese fishermen who happened to be offshore. Money and medical facilities were made available to the victims by the U.S. government. In the fall of 1954, Russia detonated a nuclear device, and radioactivity from the shot circled the globe. Following these two incidents, fallout became a household word and a matter of national and international concern for a few years. But attention focused on global fallout, not local fallout in the Southwest.

As the start of Operation Teapot approached in early 1955 at the Nevada Test Site, public relations activities intensified. Graves, AEC public relations officials, other AEC personnel, and Public Health Service experts once again fanned out among the small communities surrounding the test site.[50]

The AEC took no chances. There was a film, there were speakers, and there was a new booklet. The AEC published 55,000 copies of a pamphlet entitled *Atomic Tests in Nevada* and distributed them throughout the region and in Washington. The pamphlet, which featured Disney-like drawings and a grade-school-level text, had a rendering of a cowboy with hat tilted back, calmly sitting astride his grazing horse while an atomic explosion illuminated the early morning scene. Underneath the drawing was the notation that fallout "does not constitute a serious hazard to any living thing outside the test site." There was no mention of cancer being one of the possible effects of exposure to radiation, although radiation sickness was discussed. In 1957, when the pamphlet was updated, an AEC official noted: "The 1955 booklet was somewhat less comprehensive in discussing scientific phenomenon and offsite hazards than was desirable, but such matters were considered then 'too sensitive.'"[51]

Back in the nation's capitol, AEC chairman Strauss issued a press release on February 15 that minimized the danger of global fallout. Local fallout was mentioned only fleetingly. "The hazard has been successfully confined to the controlled area of the test site," said Strauss. The commissioners set out to dispel the fears of the public that were being voiced at that time.[52]

Besides the effects of fallout on human health, which had emerged as a major issue with the 1954 incident, the public was concerned about the use of animals in experiments at the Nevada site. The AEC received some 2,000 protests. Dogs and sheep attracted the most attention. The use of monkeys, because of the humor involved in simians "flying" drone aircraft, attracted few complaints.[53]

At a preseries press conference in Las Vegas, Elliott, Graves, and Bugher briefed reporters on the safety precautions that had been taken for Operation Teapot. Bugher, as the ranking official, recapped what the others had said and then added a familiar refrain: "Despite our concern and the amount of effort and study which has gone into this, we have not yet found a single individual whom we could honestly say has been detectably injured by any of the fallout from these Nevada operations," though it was true, he said, that "in one or two instances" the amount of radiation had exceeded the AEC standard for the offsite population.[54]

Again, the difference between what was being said publicly in

press releases, speeches, pamphlets, and films and what was discussed privately by the commissioners in the secrecy of their meeting room was remarkable. In the wake of public concern over the
first tests that year, the commission received a letter from Senator Clinton P. Anderson of New Mexico, who was chairman of the
Joint Committee on Atomic Energy. Anderson suggested that only
low-yield devices be tested at the Nevada site. The commissioners
were stunned—they considered Anderson a loyal supporter of their
program—and they took an unusual step. They ordered that the
minutes be changed from the usual summary account to a verbatim
transcript. The commission reviewed the transcript of the February
23 meeting before allowing it to stand as the official minutes. In this
manner, a remarkable conversation was preserved.

Chairman Strauss read Senator Anderson's letter, which represented an abrupt reversal in the senator's opinion about the safety
of the site. Commissioner Willard F. Libby, a chemist, said, "I am
pretty disturbed by this. I noted, Lewis, that you had cooled off
about the Nevada site."

"I had been cool long before," said Strauss. "My coolness started
in the spring of 1953, but I have never discussed this with Anderson.
This is spontaneous."

Libby said, "I think this will set the weapons program back a lot
to go to the Pacific."

"I have gone along with the majority of the commission that this
is the thing to do," said Strauss. If it was his choice alone, said the
chairman, he would have scheduled the two largest shots in the
current Nevada series for the Pacific site.

The commissioners then discussed the delay that such a shift
would have caused. General Manager Kenneth D. Nichols said, "My
immediate reaction is that we went over very thoroughly the criteria
that were set up for conducting the tests. It is my understanding
that it has been rather unusual weather out there."

Libby thought that the public furor over the tests would die down
as Operation Teapot progressed. Two small shots had already been
fired. Strauss said, "Yes, and it was a little one yesterday. But they
made as much fuss about it as if it had been a big one."

The chairman mentioned that a Nevada legislator had introduced a bill asking the AEC to leave the state, but both Las Vegas news-papers, who he said rarely agreed on an issue, cited the economic benefits of the test site and the state's contribution to national de-fense as reasons for the tests to remain where they were.

"That is a sensible view," said Libby. "People have got to learn to live with the facts of life, and part of the facts of life are fallout."

Strauss replied, "It is certainly all right, they say, if you don't live next door to it."

"Or live under it," added Nichols.

Commissioner Thomas E. Murray said that nothing should be allowed to interfere with the present series. Strauss then discussed the optimum direction for fallout to disperse from the Nevada site, and Nichols asked if east was one such direction. Strauss replied, "No. East they go over Pioche and over St. George, which they ap-parently always plaster." The chairman added, "I have always been frightened that something would happen which would set us back with the public for a long period of time."

Commissioner Murray said that he, too, was interested in the public but the Nevada site was needed because of the great distance to the Pacific site. He suggested that Anderson be mollified: "I would tell him, sure, we agree with him. We will look for a new site, and I would talk about Alaska and the northern sections. As a matter of fact, who was it that was in here a while back that had an idea to go up north instead of the Pacific?"

"Point Barrow [Alaska]," said Libby.

"Put the name of the place in the letter," directed Murray. Nichols said he would get the AEC staff working on the reply right away.[55]

At a meeting of the commission with the Military Liaison Com-mittee the next day, Strauss said that a letter was being drafted to Senator Anderson that emphasized the advantages of the Nevada site over the Pacific site. Department of Defense officials at the meet-ing said there was an urgent need for troops to participate in the tests—a need that could not be met at the Pacific site. Nichols said he strenuously objected to a suggestion by Corbin Allardice, execu-tive director of the joint committee, that Anderson's letter be made public.

Nichols got in touch with Allardice the same day and asked that he not release the letter, because the commission was being urged by Los Alamos to continue use of the site, and the controversy that might result from the letter being made public "might preclude its use." The laboratory was in Anderson's home state. The letter was not released.[56]

Nichols, a retired major general, was a good example of the type of person who headed the extensive AEC bureaucracy and typified the military dominance of that supposedly civilian agency. An engineer, he had been involved with the Manhattan Project and then had become chief of the Armed Forces Special Weapons Project (AFSWP), which was responsible for the participation of all armed forces in the development of nuclear weapons. He was appointed to that post by President Truman, who extracted a promise from General Nichols and Chairman Lilienthal, who heartily disliked each other, that they would work together. Nichols was also the army member of the Military Liaison Committee, the policy bridge between the commission and the armed forces. In 1953, at the request of President Eisenhower, Nichols resigned from the army and became general manager of the AEC.

Speed was what the former general emphasized at the Allen trial. "The main purpose of the continental test site was to speed up development of the hydrogen bomb . . . to speed up to where we would be certain that we would be ahead of Russia. . . . The overriding consideration was a necessity to proceed faster with testing . . . and it is a good thing that we did speed it up." The idea was to hold the health risks to a minimum.

Judge Jenkins asked, "Did you anticipate losing any people as a result of the radioactive fallout?"

"No, I don't think we did," said Nichols. "I don't think we ever felt we would make that bad a mistake, that we would be able to control the conditions such that there would not be serious results."[57]

Nichols left the agency in 1955 and was replaced with another former general, Kenneth E. Fields, who had a similar background. Fields was a special assistant to General Leslie R. Groves, who had headed the Manhattan Project, and had been the director of the Division of Military Applications, the most powerful of all the

AEC's divisions, for five years before he was named general manager in 1955 and had to resign his commission. (It was Fields who calmed down Congressman Stringfellow after Shot Harry.) Among other duties, the general manager represented the commission before Congress, and during one very busy period Fields testified thirty-one days—sometimes twice a day—within a period of three months.

There was a camaraderie between commissioners, the general manager, and division directors. "You knew each other," said Fields. "You were together all the time. And the division directors were considered at a very high level. We were very much together socially as well as officially. I mean, it was an atmosphere of a corporation that got along very well together." Fields left the AEC in 1958. With their long, overlapping careers in the nuclear weapons program, Fields, Nichols, and Bradbury formed a nucleus of high-ranking bureaucrats that would have been difficult, if not impossible, for any transient policymaker to dislodge, control, or redirect.[58]

The Nevada tests continued through 1955 without any serious incidents. Near the end of that year the AEC's General Advisory Committee asked Bradbury for the long-range plans of the Los Alamos laboratory. Bradbury said he foresaw five to ten years of heavy demand for weapons development. Then the man who had been at the center of the business from Trinity onward commented:

> The future beyond that point looks somewhat unrewarding. Fissionable material will go on and on being made until the efficiency of atomic weapons will become of academic interest. Everyone will ultimately have all the weapons in all the variety wanted, and the number will probably be more than the world can safely tolerate being used.

No "unscaled peak," such as the development of thermonuclear weapons, was on the horizon. What remained to be done was to enlarge what was already in existence. For that reason, said Bradbury, Los Alamos had diversified into other areas; such as developing a nuclear reactor, nuclear propulsion, and basic research. It was a way of keeping the laboratory in business. Then Bradbury posed the unthinkable: "In short, if atomic weapons are abandoned, what should Los Alamos then be in a position to do? How should we plan today?" Every succeeding director has pondered that question.[59]

There were no continental tests of any consequence in 1956, that activity having been shifted to the Pacific, but there was the problem of readjusting the radiation safety standard for the test site, which caused a great deal of consternation within the AEC. The National Academy of Science issued a report on fallout that year that recommended a more restrictive standard than what was in force. The NAS standard of 10 roentgens in a 30-year period was not "operationally feasible," the AEC determined, so with Gordon Dunning of the Division of Biology and Medicine again the number cruncher, the AEC came up with a compromise. The goal was to find a figure that would permit testing in an unimpeded manner yet make it look like the AEC was conforming to the spirit of the NAS study.

Dunning recommended a standard of 10 roentgens in a 10-year period to the commissioners at their November 14 meeting. Strauss repeated his wish to transfer all tests to the Pacific, stating that the commission would not then be faced with such problems as establishing offsite radiation standards. Libby proposed a standard of 5 roentgens in a 5-year period. The minutes noted: "Dr. Dunning said that he believed it was much easier to announce a 10-roentgen level and to point out later that this level had not been exceeded rather than to explain why a more stringent 5-roentgen level, if adopted, had been exceeded." The 10/10 standard was adopted by the commission but not publicly announced.[60]

One week after receiving a sketchy description of the coming series from Strauss, President Eisenhower routinely approved it. Strauss promised that steps were being taken to minimize local fallout. For the Plumbbob series in 1957, a record number of thirty planned shots caused concern within the AEC. Officials did not want to alarm the public and so held off on an announcement of the total number of shots.[61]

With a new series approaching, public relations once again came to the fore. Elliott was busy. He heard from Gladwin Hill of the *New York Times*, and passed on to his superiors, the fact that Hill and an Associated Press reporter had surveyed public opinion in the region to learn "if political discussion had aroused concern over Nevada tests and fallout. He said they found nothing on which to hang a story, most people saying that they had been through this before, so why get excited now?"[62] Again, the emphasis was on being reactive, not informative. When reviewing the public information plan for Plumbbob, Commissioner Libby commented, "The main thing in this paper is that you take care of public relations in case of an unfortunate or unusual development or accident." Shot Harry was still very much on the commissioners' minds.[63]

In April, one month before the Plumbbob series was to kick off, Elliott outlined how he planned to handle the preseries press conference. "Graves or Reeves [the test manager] will say in May 13 briefing that number will be highest for any Nevada series, but will quickly add that fallout will be less, and why." Graves and Elliott wanted to push the fact that small weapons were being tested. They suggested that the shot that was to be open to the press be one from a moored balloon "because of obvious interest and need to sell their safety."[64]

A problem with the Plumbbob series was not long in coming. In fact, it occurred on the first shot. Boltzmann was a 12-kiloton device detonated atop a tower on May 28. Brigadier General Alfred D. Starbird, director of the AEC's Division of Military Applications, reported to General Manager Fields: "The local fallout produced was considerably greater than predicted. Rain showers and loose earth at the base of the tower were suspected as possible causes of this unpredicted fallout." There was also another cause. It was Commissioner Libby's idea to load the cab, which sat atop the 700-foot tower and held the device, with sand so that the fallout, being heavier, would descend quicker—hopefully, within the test site. It did descend quickly, but not quickly enough.[65]

Boltzmann also vividly demonstrated the phenomenon of a hot spot—a small area of much greater radioactivity than was measured in the surrounding region. The Boltzmann hot spot, calculated at the time to be seven times more radioactive than the surrounding region, was thought to have been caused by wind passing over a mountain and depositing fallout on the leeward side. Rain could also cause a hot spot. The most intense area of radiation from Boltzmann was located just north of the test site along the state highway.[66]

By the time of Operation Plumbbob, offsite radiation monitors from the Public Health Service, who had replaced army monitors in 1953, were living in the communities they were responsible for and were encouraged by the AEC to become part of everyday life. They carried extra cans of gas to help stranded motorists. "Compliments of the AEC," they said. This was the public relations aspect of their jobs. In September 1957, monitors discussed with ranchers and miners north of the test site the possibility that they might have to be evacuated for certain shots. They would be reimbursed for costs and given $10 or $12 a day. There was no mention of who would do the chores in their absence.

The monitors made regular stops at the Fallini ranch on State Highway 25. After the fallout from Boltzmann and other shots that summer, Helen Fallini told her government visitor, "The AEC does not give a damn about the ranchers."[67] Mrs. Fallini and other residents of the area were bitter. The previous year, her nephew, Martin Bordoli, had died of leukemia. His was the first cancer death that was publicly linked to the tests. The petition and articles that resulted were the first concerted public questioning of the tests in terms of human health. Did their impact justify the deceptions carried out in the name of national security? That was the question these activities posed in hindsight.

Martin, age seven, lived five days a week with the Fallinis while he attended the small school at their Twin Springs Ranch. The Bordoli ranch was thirty-seven miles up a dirt road into the mountains. Radioactive clouds frequently enveloped both ranches. Bordoli Creek meandered past the ranch house and outbuildings shaded by cottonwood trees in an isolated valley cupped by mountains. From the ranch Martha Bordoli, the sister of Mrs. Fallini, saw the flashes of light and the fireballs rise from the nearby test site. Shortly thereafter the earth shook and then, depending on the strength and direction of the wind, a cloud passed over the ranch. Sometimes the cloud had a pink glow, and the air smelled funny. Sometimes it was black and rolled in like ground fog.[68]

The Bordolis had three children. They complained of headaches. The reddening of their skin did not resemble a sunburn. There were small water blisters atop the burned areas. Mrs. Bordoli bathed her children and coated them with Noxema. The family ate the food that they raised or that was available in the surrounding area, including plentiful amounts of fresh venison. The doctor who treated young Martin said the radioactive fallout may have caused his leukemia, which also strikes children under ordinary circumstances.[69]

Mrs. Bordoli received no visits from AEC personnel at the ranch prior to Martin's death, nor was the family given any warning or told what precautions to take. After Martin died, two representatives of the AEC visited her at the Fallini ranch. "It was a little P.R.," said Mrs. Bordoli. "You know, talking to the people. He was saying that the fallout would not affect us that much. What little we got. So he said he had children. I told him that I would gladly take his children to our ranch and baby sit them for the rest of the summer while they were setting off those bombs. He said, 'My God, woman, don't wish that upon me.'"[70]

The mother circulated a petition among her far-flung neighbors, and seventy-five of them signed it. She sent the petition to elected and appointed officials in Washington. It asked that the tests be suspended "or that some equally positive action be taken to safeguard us and our families." The signers of the petition described themselves thusly: "We are not excitable or imaginative people, most of us coming from rugged ranch families, but neither are we without deep feelings for each other and our children."

The responses were not long in coming. In the fall of 1957, Nevada senator George W. Malone, pointing out that some scientists, such as Linus Pauling, had created concern over global fallout in the newspapers, wrote that "The President has questioned these reports coming from a minority of scientists, some admittedly unqualified to comment on nuclear testing, and he has said, it is not impossible to suppose that some of the 'scare' stories are Communist inspired." Actually, Senator Malone had badly misrepresented

President Eisenhower's comment. Eisenhower had said at a press conference a few months earlier that the scientific criticism looked like "an organized affair," and then, in reply to a question, said he did not mean to infer there was a conspiracy or "a wicked organization," just an honest disagreement by a group of scientists who had no specific expertise in fallout matters. But Mrs. Bordoli did not know this.

AEC chairman Strauss, who had said privately two years earlier that fallout was all right as long as "you don't live next door to it," wrote the distraught mother that the risks were "exceedingly small in fact when compared to other risks that we routinely accept every day." Strauss said he had a wife, a son, and three small grandchildren and believed the risk of fallout affecting health was much smaller than the danger of nuclear war, which the testing deterred. Strauss later died of leukemia.

In preparation for a possible statement from the AEC on the boy's death, Gordon Dunning estimated that Martin had been exposed to slightly less than 1 roentgen from all the tests, an amount he said a youngster would receive from natural sources. Since there were no monitors at the Bordoli ranch to take readings and the family had not worn film badges, this was only a guess. Dunning also denigrated Mrs. Bardoli's symptoms of radiation sickness, stating that the time frame for her loss of hair did not fit the regular pattern and that cessation of fingernail growth was not confined to such a cause.

Mrs. Bordoli felt defeated by these and similar responses to the petition. She felt misunderstood and betrayed by her government and its representatives. She said later, "I believe that we were just feeding our children and families poison from these bombs. At least I did let the government know at that time how I felt about the whole thing." The Bordolis sold the ranch at a fraction of its value and moved away in 1958.[71]

The nearest town to the Bordoli ranch was Tonopah, some sixty-five miles west of the Fallini ranch. The only repeated questioning of the AEC assurances of safety during the years of atmospheric testing was voiced by Robert A. Crandall, the feisty editor of the *Tonopah Times-Bonanza*. Crandall took up the cause of Martin Bordoli's death. He pointed out that the Bordolis lived in an area of heavy fallout.[72]

Crandall got in touch with Pauling, a Nobel Prize winning chemist who was pointing out the dangers of global fallout at the time. Pauling later won the Nobel Peace Prize for these efforts. They talked on the telephone, and Crandall quoted Pauling in his newspaper. A somewhat deceptive front-page headline over a story featuring the comments of Pauling read, "Local Citizens 'Give Up' 1,000 Years." This headline was based on Pauling's comment that the tests may have shortened residents' lifespans by an average of three months. Crandall then added up the area's population and did some quick multiplication. He used exaggeration to make his point, a technique used almost a century earlier by Samuel Clemens (later known as Mark Twain), who wrote for the Virginia City newspaper not too far to the north. Later, the AEC's Richard Elliott, who was having a difficult time with Crandall, issued a statement. Crandall headlined that story "AEC Official Contends Fallout Here 'Slight.'"[73]

The editor then went on to ask a basic question: Why was it that fallout was not good for Las Vegas, which the AEC attempted to miss, but all right for those less-populated, but nevertheless inhabited regions north of the test site? Crandall said he "just got riled up because he, his neighbors, and their kids suddenly attained the status of 'virtual uninhabitants,'" a play on the phrase "virtually uninhabited desert terrain," used by the AEC in its press releases to describe the region north of the test site into which fallout from the largest number of shots descended.[74]

Using the argot of Nevada, he ran this question-and-answer sequence in his newspaper:

Q) Aren't we folks who live around "uninhabited areas" and who like to go hunting, fishing, camping and prospecting in out-of-the-way spots, betting the short end of the odds when you refuse to fire an atomic shot when the wind blows away from us?

A) And then, inevitably as it must when the AEC finds itself on crumbling ground, it turns questioner to demand: "Do you want us to halt the Nevada tests, and let our enemies win the atomic race for survival?"

Q) No, we don't say that at all. But what we do say is this: "All right, AEC, go ahead with your tests. But if you must shoot craps with destiny, first throw away the loaded dice."[75]

Crandall and his wife, Minette, became thorns in the side of the AEC. They could not be won over. William P. Becko, who was frequently at the test site in his role as district attorney for Nye County, said, "They waited each week for Crandall's newspaper to come out and shuddered at the coverage." Washington asked Elliott to take "eductional and corrective action" with the Crandalls. Elliott sought to reassure his superiors in Washington. He wrote, "We feel from our contact with Tonopah leaders during the February meeting that the attitude of the editor and his wife is not that of their community. Our monitors endorse this opinion as a result of conversations with many community leaders. We will continue our efforts."[76]

One of the AEC blandishments that the Crandalls repeatedly turned down was a tour of the test site and a chance to witness a shot. Crandall was terrified of fallout, and on days when a shot had previously been announced, he remained home and sent Minette to the office. They remained in contact by telephone. Becko, a friend of the Crandalls, said Minette later died of cancer in Chicago and Robert was killed while witnessing a robbery in San Diego.[77]

It was through Pauling that Paul Jacobs, who was working for the *Reporter* magazine, learned of Crandall in early 1957. The journalist, who described himself as a social critic, interviewed Crandall, the Bordolis, the Fallinis, and other ranch families north of the test site. Jacobs then went to Las Vegas, where he met with an unnamed Public Health Service official who said he had a classified report in his desk that might be of interest. The official left his office. Jacobs locked the door, searched the desk, and found the report, stamped "Official Use Only." He slipped it into his briefcase and walked out of the office, worried that he might be caught and sent to jail. The journalist later recalled in a television documentary that to his "astonishment and disgust and horror" what Crandall and the offsite residents alleged was verified in the PHS report on the 1953 Upshot-Knothole test series.[78]

After completing his research in Nevada, Jacobs visited AEC officials in the Washington area in February. He talked with people in the divisions of Information Services and Biology and Medicine. After a meeting with Dunning and others his views were modified, at least according to an AEC staff report. Jacobs then went to Albuquerque to talk to Elliott. The AEC staff monitored his activities closely. Summations of interviews went right to the top.[79]

The resulting story, "Fallout From Nevada," was published in the May 16, 1957, issue of the *Reporter*. One week before, Strauss had warned fellow AEC commissioners that it would appear, and the AEC staff, which got advance copies, started drafting a reply even before the magazine hit the newsstands. Selby Thompson, the acting director of Information Services, told the commissioners, "This article is almost certain to arouse alarm among the residents of the area, particularly since it comes within a few days of the beginning of Operation Plumbbob."[80] Actually, the story caused very little alarm, although it was remarkably prescient.

The lengthy article was the first probing account in a national publication of the weapons tests and their possible health effects. Jacobs wrote, "The AEC seems to have been shifting from one foot to the other, torn between its belief that weapons testing is one aspect of the rich possibilities of the atomic age and its commitment to the protection of the public health. Of course, most of the AEC staff firmly believe that there is little danger from the kind of weapons testing carried out in Nevada. But a man's conviction that the tests are safe might help to rationalize his belief that they are valuable, or his conviction that they are valuable might help to rationalize his belief that they are safe."[81]

The magazine received an award for the story, and Jacobs was given a bonus from editor-publisher Max Ascoli and two tickets to the Broadway musical *My Fair Lady*. The AEC issued clarifications and denials, further questions from Jacobs were answered in writing, and interviews were taped. Jacobs repeated his effort for the *Atlantic* in the February 1971 issue, this time widening his inquiry to the safety of nuclear power plants and lung cancer among uranium miners. This time the AEC commissioners knew the previous August that Jacobs was preparing a critical article.[82]

The writer was disappointed in the public response to both articles, but the response within the AEC was quite remarkable. In January 1970, Jacobs appeared in a segment of the public television show called the "Great American Dream Machine" that was critical of the AEC's nuclear energy program. Report upon report was written within the bowels of the AEC on the television program and the magazine article, and the reports rose to the top, where they were condensed into a press release and letters of complaint to the editor of the *Atlantic* and the president of National Educational Television. It was a massive effort at nit-picking.[83]

Paul Jacobs, like Dale Haralson, was a victim of the disease he was so concerned about in others. A light smoker, the San Francisco Bay area journalist died of lung cancer in 1978. Jacobs and his friends believed that he had contracted the disease while researching his stories. A more cautious Pauling said that the cause could not be determined with absolute certainty.[84]

The next year, public television broadcast the documentary "Paul Jacobs and the Nuclear Gang." It had been shot the previous year. The circle had closed. One scene showed Jacobs interviewing a dying Paul Cooper, the army veteran who had witnessed Shot Smoky. Bill Curry wrote a review of the documentary in the *Washington Post*, declaring that "it is impossible to forget that the experts criticized Paul Jacobs twenty-two years ago. Said he jumped to conclusions, didn't know what he was talking about. They recited their knowledge and credentials, gave us their assurances and crossed their hearts and hoped to die that everything was okay."[85]

The Plumbbob series of tests continued in an uninterrupted manner despite the public outcries from Mrs. Bordoli, Crandall, and Jacobs. Shot Smoky, about which there was some advance concern, was detonated atop a 700-foot steel tower in August. It was rated "a very important one development-wise." The reason Smoky lived up to its name was that sixty tons of coal was placed at the 100-foot level of the tower, and large amounts of lead and paraffin were loaded into the cab at the apex of the tower. All were vaporized. It was a Livermore laboratory shot.[86]

Over two thousand Canadian and American troops maneuvered through the area in a test of the army's new pentomic battle formations. Forty aircraft were involved. An AEC press release stated, "Army officers in charge of the operation said they were very pleased with the manner in which the exercise was conducted, and the speed with which it was executed." Three hundred mice were placed in bomb shelters being tested for the French and German governments.[87]

Kermit Larson of UCLA traced fallout from Smoky to Rock Springs, Wyoming, where he detected a hot spot in the surrounding countryside after a thunderstorm. The scientist returned there a year later and found that the strontium 90 in the bone marrow of rabbits equaled the amount of the same radioactive isotope found in rabbits two miles from ground zero. This finding, said Larson, "sort of shook us up."[88]

The UN General Assembly was to meet the following month and, with a half-dozen tests still to be made in the series, General Starbird of the Division of Military Applications sent a memo out to all AEC facilities involved in the tests instructing them to put as little information out as possible during the disarmament debate. The idea was not to disturb the international waters any more than they had been. There was to be a low-key termination announcement on Plumbbob and no further elaboration on the Hardtack series coming up the following year.[89]

The AEC embarked on an orgy of atmospheric tests in 1958, rightly suspecting that some limit would be placed on them before the end of the year. The public debate over global fallout was reaching a crescendo. The Russians, after completing a series of tests, announced a suspension. President Eisenhower favored a ban on further tests, while Strauss opposed it. The two superpowers moved toward a voluntary suspension of tests.

Operation Hardtack II, nicknamed Operation Deadline by the press, got underway at the Nevada site on September 12. The background information for the series issued by Elliott's office stated, "The only known cause of offsite injury to a human as a result of Nevada nuclear testing came during the 1957 series when a man in Hiko, Nevada, strained his neck from a sudden movement when a blast wave from a Yucca Flat shot reached his home and broke a window."[90]

The term *known* apparently meant publicly admitted by the AEC or legally determined by a court. A number of possible cases of radiation sickness had been noted by the AEC, which refused to recognize any of them as genuine. Formal claims and lawsuits were equally ineffective. The claims of two California prospectors, Owen D. Atkins and Clarence F. Terry of San Bernardino, who were in the Virgin Mountains southwest of the test site during Shot Simon, were settled for about $400 apiece. They said their emphysema had been exacerbated by the fallout. A number of cancer suits had been filed against the AEC, including one in 1955 by Martha Sheahan, wife of the owner of the Groom Mine. At least two others were filed that year. By 1958 about a dozen lawsuits had been filed against the AEC. As far as is known, none were successful.[91]

As the midnight October 30 deadline for a voluntary moratorium on further testing approached, the tempo of tests increased markedly. Within a month and a half, thirty-seven tests were conducted at the test site. There were four tests on October 22, two on the twenty-fourth, three on the twenty-sixth, one on the twenty-seventh, three on the twenty-ninth, and four on the thirtieth. There was fallout over Los Angeles during the last days, and Mayor Norris Poulson called the White House to complain.[92]

The era of atmospheric testing, which effectively came to a close with the moratorium, ended with a whimper. The last shot was scheduled, then postponed, then scheduled and postponed again and again. The problem was potential blast damage to local structures. One of the offsite monitors recalled:

The blast was predicted at one time to hit Ash Meadows, which was a well-known cathouse. And about every thirty minutes or so I would call up the madam and ask her to please open her windows [to lessen such damage]. She was very nice about it. But I couldn't help thinking, boy, it was cold and bad for business. It went on and on and we never did get that shot off. I finally called her up at midnight and said, "Well, I am not going to bother you anymore." And she said, "That's fine. Why don't you come on over."

The test was canceled by President Eisenhower, who did not want to violate the midnight deadline.[93]

The Soviet Union broke the voluntary moratorium in September 1961 with the resumption of atmospheric explosions, and the United States followed shortly thereafter with a series of underground blasts in Nevada and atmospheric tests at the Pacific site in 1962. More underground tests were conducted at the Nevada site in 1962. There were problems with offsite radiation that summer.

Sedan was a 100-kiloton thermonuclear device buried 635 feet underground. The event, as shots were now called, was designed as a Plowshare Program experiment in earth-moving techniques. Plowshare had peaceful purposes, such as digging canals, as well as military benefits. When Sedan was detonated on July 6, it blew a 1,200-foot-wide, 320-foot-deep crater in the floor of the desert that to this day is the primary sightseeing attraction at the test site. The explosion was a spectacular display. Columns of dirt, stone, and dust—a total of 7.5 million cubic yards of material—cascaded into the air. The resulting dust cloud, which reached a height of 12,000 feet, was 50 percent greater than predicted. At the direction of the Kennedy White House, no pictures of Sedan were released.

The roiling cloud of radioactive dust, which reached from the ground to the 10,000-foot level, crept up the desert valleys. It enveloped thirty beagles in wire cages that had been placed twelve, thirty, and forty miles from ground zero. Their mouths were taped shut to prevent them from ingesting fallout. It was a test of an inhaled dose. Four died immediately, three within twenty-four hours, and three within seventy-two hours.

Some of the inhabitants north of the test site were evacuated, others were not. "It seems reasonable to conclude that none of the offsite population in the vicinity of the Project Sedan test site received significant amounts of radiation either internally or externally," stated the project manager's report of the shot. A postshot press release stated that most of the radioactivity was trapped underground, although the dogs may not have been aware of that fact.

The cloud engulfed the Fallini and Bordoli ranches and other isolated habitations. The streetlights in Ely were turned on at 4 P.M. that summer afternoon. The cloud then passed over Salt Lake City, where fallout from Sedan and two other shots deposited radioactivity in the milk supply via the feed that cows ate, and it then headed directly east from the center of Wyoming, dipped just south of Chicago and passed out into the Atlantic Ocean between Delaware and North Carolina. A scientist from the sponsoring Lawrence Radiation Laboratory said, "Sedan was clean, but not clean enough." The technique, it was determined, was sufficiently workable to dig a sea-level canal.[94]

In the summer of 1963 the United States and Russia agreed to ban atmospheric tests, and since then tests by the two countries have been underground. Between the signing of the Limited Test Ban Treaty on August 5, 1963, and the end of 1970, small amounts of radiation were detected offsite following twenty-six events, according to test site officials.[95] Three of those events were of special interest, as they indicated that the AEC was still not particularly forthcoming.

On Friday, March 13, 1964, the Pike Event was fired 390 feet below ground. A dense, black radioactive cloud vented through a fissure in the earth for about one minute and then drifted south over Las Vegas and was tracked to the Mexican border, where it presumably crossed into that country. Tracking aircraft turned back at that point, and Mexico had no radiation detection devices in the border area.[96]

The AEC made no announcement of the incident, as there was no inquiry on the matter from the press that day. No questions, no announcement was the AEC policy, which had a bit of Catch-22 in it, since how would a newsman, who was not normally armed with a radiation detector, know when radiation was blown offsite within that vast region?

The next day a Las Vegas newspaper reporter called the test organization. Offsite radiation monitors were seen on the Strip. What about it? The AEC, after internal debate, issued a statement. The radiation readings were minimal, said the one-paragraph statement, which was designed to be read over the telephone to the press. Under the notation "Following for use if inquiry made" were two additional lines that read, "This and other such incidents involve such low levels of radioactivity that the test ban treaty would not be affected." Who would have known to ask the key question? A treaty violation consisted of "radioactive debris" crossing an international border. Robert L. Brown, editor of the *Las Vegas Review-Journal*, apparently asked the right question, as he quoted the answer in his story.[97]

The two additional lines raised an interesting question: What were the "other such incidents"? James E. Reeves, the manager of the Nevada Operations Office, wrote in a classified memorandum on the Pike incident that there had been "a dozen or so" other venting incidents. He also noted that "the backlog of goodwill and good relations with the public of the NTS region during our days of open operations in the middle and late 1950s now is dwindling because of our restrictive public information policy."[98]

Three days after the venting incident, the AEC issued a press release stating that the Public Health Service had been given money for a quick study of radioiodine in the Las Vegas milk supply. Small amounts of the isotope had been found in the milk, but the amounts were far below the hazardous level.[99]

The debate that went on secretly within the AEC and the Johnson administration following the Pike event, according to Oliver Placak, who was the PHS official in charge of the offsite monitors, revolved around the question "Should they inform the world that they had violated the test ban treaty or should they just forget the whole damn thing?" Morgan Seal, another monitor, said at the offsite monitors' conference, "Well, to this day that offsite Pike report has never been released."[100]

After the shot the draft PHS report on Pike was quickly gathered up and stamped "secret" on each page by the AEC classification officer. Gordon Dunning of the Washington office then stepped in to review it, which angered PHS personnel at the test site, who thought this was a case of public data being suppressed. In addition, Dunning told them that the report had been reviewed at the White House and that the decision had been made not to issue it in its present form. There was nothing the PHS officers could do, so they wrote memos for the files "so if the question ever arises of suppressing data, or there are any complaints through channels about not getting out reports, the facts as they occurred will be readily available." Placak thought that there might be repercussions at some later date.[101]

On April 2 the AEC assured President Johnson through McGeorge Bundy, Special Assistant to the President for National Security Affairs, that for the next series of tests, for which approval was being requested, precautions would be taken to preclude a repeat of the Pike incident. "We remain confident that we can maintain an active underground weapons testing program without creating undue risk of violation of the Limited Test Ban Treaty," said acting AEC chairman John G. Palfrey.[102]

The commission approved Dunning's sanitized report on Pike and authorized its publication. A paper by several PHS scientists on the Pike event was submitted to the AEC commissioners for approval for publication. An AEC staff report, which had Dunning's touch, noted: "Although the paper does not discuss wind trajectories, correlation of the Los Angeles readings with wind trajectories that are in the public domain could be taken to show that the material passed through Mexico on its way to Los Angeles." Such information, the report stated, could "provide a possible basis for charges that the United States has violated the test-ban treaty." Dunning, a fallout expert, had provided the technical input on the trajectories. The commission classified the paper secret, and there it lay.[103]

There has been at least one other known American violation of the 1963 treaty. Radioactive fallout from Schooner, a 30-kiloton Plowshare experiment detonated on December 8, 1968, drifted to the northeast. A month later there were media reports, but no

AEC announcement, of radiation from Schooner being detected in Canada.[104] Undoubtedly there have been other violations.

Schooner also kicked up a fuss closer to the test site. Utah Governor Calvin L. Rampton, after listening to medical experts discuss venting incidents, said he thought that Utah was better off without the tests. He was joined by a host of others in denouncing the tests. They included Howard Hughes's Nevada-based organization; Paul Schrade, a liberal Los Angeles labor union leader; Utah and Nevada academics; and representatives of the National Committee for a Sane Nuclear Policy (SANE) and the Peace and Freedom party. As usual, opinion was mixed. Clifford A. Dean, Jr., a chemistry professor at Dixie college in St. George, said, "People who don't understand the testing are bothered. Many of them are professional 'aginners.' We, who know something about it, are not bothered." The AEC gave Governor Rampton the customary assurances and said that testing would continue.[105]

There was no mention of the Pike or Schooner incidents when the Reagan administration released a report in October 1984 on Soviet violations of the 1963 Limited Test Ban Treaty and other nuclear weapons agreements. The report pointed out that underground tests conducted by the Russians had leaked radioactive debris that had been detected beyond Soviet borders. The Soviets replied that the report was an attempt "to ascribe to the other side its own misdeeds."[106]

One week before Christmas in 1970 a 10-kiloton nuclear device, named Baneberry, was detonated at a depth of 912 feet. Three minutes later a black cloud, which resembled smoke from burning tires, issued from a 315-foot-long fissure that suddenly split the floor of the desert. The cloud engulfed the nearby Area 12 Camp, which housed some 900 civilian test site workers. Because the fire chief could not immediately be found, the order to evacuate the area came too late. A security guard, Harley Roberts, ran from trailer to trailer warning the occupants. William Nunamaker was working in a tunnel at the time. The black cloud moved slowly off to the north. Radioiodine was found in milk as far away as Bakersfield, California, and Mount Pleasant, Utah.[107]

Once again the AEC was slow to release information, and when it did it was incomplete. Four days later Congressman Chester Holifield, chairman of the Joint Committee on Atomic Energy, angrily wrote AEC chairman Glenn T. Seaborg: "This is another example of a continuing series of events concerning nuclear health and safety matters about which I learned initial details from those outside the AEC." The instructions regarding press information on Baneberry issued that same day from AEC headquarters directed Nevada personnel to "stress specifics with respect to absence of health hazards."[108]

Roberts and Nunamaker died of leukemia, and their widows filed a suit under the Federal Tort Claims Act. After conducting the trial in early 1979, Judge Roger Foley of the federal district court in Las Vegas issued his decision in two parts. The first part, issued three months before the start of the Allen trial in 1982, stated that the government was negligent in its evacuation and decontamination procedures. The second part, handed down two months after the end of the Allen trial, held that the radiation exposures were not sufficient to cause leukemia. Judge Foley accepted the government's dose estimates. The plaintiffs proved negligence but not causation. They lost. The government had kept its record intact.[109]

CHAPTER 7

THE VICTIMS

THE DOWNWIND region is a separate place, cut off from the rest of the country by deep gouges in the earth, giant upthrusts, and desolation. To the west is the trench of Death Valley and the crest of the Sierra Nevada. The low, hot desert incised by the Grand Canyon lies to the south. East is the Colorado Plateau and its maze of canyonlands. There are more mountains to the north, and a long trail of misery traversed by the cancer patients on their way to and from the doctors and hospitals in Salt Lake City. The isolation and harshness of the surroundings demand self-sufficiency and belief—the two primary characteristics of its inhabitants. With the coming of the atomic tests another element was added to the delicately balanced equation of life and death in a harsh environment.

The course to follow through this region of grief thirty years after Shot Harry was from west to east, the direction of the prevailing winds. The wind was quite noticeable that spring, and the desert wildflowers were gorgeous. Nevada was rock-hard and coarse. Tonopah differed markedly from the orderly Mormon-dominated hamlets to the east. It was what was euphemistically referred to as a rough-and-tumble western mining town. Supposedly a prospector picked up a rock to throw at his recalcitrant burro and discovered silver in 1900. The rush was on, and Tonopah became an instant town in the high desert. Along with the boom came the Mizpah Hotel, a symbol of elegance where Jack Dempsey once worked as a bouncer. In 1953 the big news was the lifting of the eleven-year ban on prostitution. The *Times-Bonanza* noted daintily, "The first establishment to begin taking on temporary lodgers will probably be the Nugget bar where Alice Nashlund was fatally beaten in December, 1951. Her murder has never been solved."

Three decades later the lead story in the newspaper, whose motto was "For a Greater Nevada," concerned the opposition to the location of a high-level radioactive waste repository at the nearby test site. "The theme that was repeated over and over was that Nevada had done its share with nuclear energy," the newspaper stated. Had editor Crandall still owned the paper, he would undoubtedly have spearheaded the opposition, which came mostly from Reno and Las Vegas residents.[1] East from Tonopah the landscape rose and fell with the regularity of the ocean. The north-south trending valleys and mountains were classic basin and range country. The state highway that traversed the north end of the test site was a lonely, repetitive road. It wound up long grades to the Queen City, Coyote, and Hancock summits, then slid down to the dry basins. At the summit were piñon and juniper forests. There was sagebrush at the midway point, and creosote bushes and Joshua trees fringed the desert playas, as they did at Yucca and Frenchman flats to the south. The brisk spring wind pushed dust devils across the dry lake beds.

Ranching was the predominant occupation. The Fallini family owned or leased the land for twenty-five miles in one direction and fifty in another along this road. Their most vexing problem in the early 1980s was the proliferating wild horses, which were competing with the cattle for feed. Most ranches, however, had less than 350 head of cattle and a much smaller land base. The vast majority of ranchers had full or part-time jobs elsewhere to supplement their meager agricultural incomes. Their average age was about fifty, they had all lived in the region about the same length of time, and the overwhelming majority came from families who had also ranched. In other words, things did not change here very much over the years. The Bureau of Land Management, a federal agency, took an informal survey and came up with these results:

> The positive value placed on the small size of local communities, the positive aspects of a rural atmosphere, the appeal of clean air and moderate weather, the easy access to outdoor rec-

reation, the feeling of friendliness and sociability; the opportunities afforded of doing things as a family; and the belief in the natural order of things (particularly the belief that change will proceed modestly and gradually without altering the county's rural character) are the values that seem to be consistently articulated by county residents.[2]

A mine was spotted here and there amidst the rangeland. The Lincoln Mine, now owned by Union Carbide, was closed because of the vagaries of the tungsten market. During the years of atmospheric testing, when it had been owned by K. C. Li, the brother-in-law of Chiang Kai-shek, the mine had been one of the nation's biggest producers of tungsten, a strategic metal that the American Tungsten Association called "one of the critical metals in the Atomic Age." Because of its importance, the mining of tungsten was subsidized by the government until a surplus developed in the late 1950s.[3]

The first relief from the unrelenting rhythm of the dry land north of the test site was the green Pahranagat Valley. There were no advance indications that a well-watered oasis existed on the other side of Hancock Summit. It just occurred, pleasantly. The valley was the first Mormon outpost west of Utah. It represented agricultural stability, as opposed to the vagaries of the mining communities.

The Whipple home, an adobe structure nestled amidst poplar trees, was originally built in the late 1860s by the owner of a silver mill in Hiko, located at the north end of the lush valley. After silver was discovered in the surrounding mountains, there was a brief boom, during which the Paiute Indians were displaced by the whites. When the ore gave out, the miners and their camp followers moved elsewhere. The water flowing constantly from the three springs, the warm climate, the good grass, and the seclusion attracted rustlers and murderers, but once the lawless elements were driven out, the Mormons took over with their agricultural ways. To those on the way to settle the valley, the Mormon leader Brigham Young said, "Set one hundred men to mining and ten to farming, and at the end of ten years the ten will be worth more than the one hundred, and probably have to feed them gratuitously."

Very little of what went on elsewhere in the world intruded on the valley until the atomic tests began. During the late 1970s, the remote valley seemed to be poised once again on the edge of the nuclear precipice. The Carter administration planned to locate the MX missile system throughout this region, but neither the missile system nor its vast impedimenta were welcome. Too many people in the valley remembered the dark clouds. When the Mormon church opposed placement of the system here, the missiles went elsewhere.[4]

Between the Pahranagat Valley and St. George there was a roadside historical marker, the only reminder of an earlier bloody stain upon the land: the Mountain Meadows Massacre, in which approximately 120 non-Mormon emigrants—men, women, and children— were brutally murdered by Indians and Mormons in 1857. It was a time of war hysteria not unlike the cold war years following World War II. The Mormons feared that federal troops were about to invade Utah. The emigrants, who were strangers, were the unwitting victims of the war hysteria, and they were ambushed. One man, Indian agent John D. Lee, was caught, convicted, and shot twenty years later for a crime in which a number of other prominent Mormons were also involved. Lee's great-grandson, Stewart Lee Udall, was one of the lawyers for the plaintiffs in the Allen case. Many of Lee's descendants were scattered throughout the downwind region, and some conceivably were victims of the latest federal intrusion.[5]

There were more circles within circles in this closed land. There was no historical marker at the turnoff to Snow Canyon, not far from the outskirts of St. George. The road descended into a red-walled canyon that was a state park and a pleasant place to camp for the night. It was in this canyon that the film *The Conquerors* was shot not too long after the fallout from Shot Harry had descended upon the region. The film, directed by Dick Powell, starred John Wayne as Genghis Khan, and Susan Hayward and Agnes Moorehead also appeared in it. Large portable fans whipped up the sand to make it look like the wild steppes of central Asia, across which the Mongol hordes rode. The fans may also have stirred up radioactive particles, a phenomenon known as resuspension. The movie's stars and director later died of cancer, as did an unusually large number of those who were on the film crew.[6]

The Hollywood connection continued. *The Conquerors* was produced by Howard Hughes, who married actress Jean Peters in Tonopah's Mizpah Hotel. In the late 1950s an employee of Hughes, who owned Trans World Airlines, asked an AEC commissioner how fallout affected weather conditions. The airline was studying the possibility that the altered weather caused problems for its flights. The Hughes assistant, Nadine Henley, asked Commissioner Libby for data from Operation Plumbbob, and Libby said he would send it.

The increasingly isolated and disease-conscious Hughes, who was acquiring a small empire in and around Las Vegas in the 1960s, mounted a vigorous campaign against the tests, which became the billionaire's greatest obsession. In the end the richest man in America could not stop the momentum of the AEC. Hughes, desperate because of concern over his business investments in Nevada and paranoid about his health, attempted to bribe Presidents Johnson and Nixon, Vice-President Hubert Humphrey, lesser politicians, bureaucrats, and scientists—with no success. The tests continued. The recluse frequently called Nevada governor Paul Laxalt to complain about the underground tests. "He was on the phone every day, saying if you continue this, you are going to contaminate your state forever," said Laxalt.[7]

The Hughes connection continued. The attorney for a former Hughes employee by the name of John H. Meier, an unsavory swindler who was in jail, contacted government attorneys during the Allen trial to see if they were interested in a fallout study done by the Hughes organization. Meier had access to Hughes's documents after his death in Mexico. The government lawyers were not interested, nor were the plaintiffs' lawyers. By coincidence, Judge Jenkins had sentenced Meier three years earlier for obstructing justice in a case relating to his obtaining overpriced mining claims for Hughes.[8]

Beyond Snow Canyon lay the Virgin River valley and St. George. The valley marked the transition zone between the basin-and-range province and the Colorado Plateau. The difference was color: the lack of it to the west, and the richness of it to the east. This was "Color Country," the term used in promotional literature. There was the red hill (Navajo sandstone) just north of St. George, and the black hill (volcanic rock) to the west. Town residents watched the atomic tests from these hills and others closer to the test site.

The town was the spiritual and mercantile center for the area, called Dixie because of its southernness. The climate was warm. Cotton and silk had been raised there in the early days; and the Mormon leader, Brigham Young, had spent his winters in St. George. In 1871 Young decreed that a temple be built. It was completed five years later. The Mormons accomplished things. Concerning the town's centennial celebration in 1961, a history by the Washington County chapter of the Daughters of Utah Pioneers noted, "Every heart was overflowing with pride and gratitude for our stalwart pioneers who made possible our abundant life in this beautiful valley we call home." Retirees began moving into the area during the 1970s. By 1980 the population had increased 82 percent over the 1970 census figure, and concentrations of selenium, arsenic, and radioactive elements that exceeded recommended irrigation water standards were found in the nearby Virgin River.[9]

The wind picked up over the unobstructed Arizona Strip, which lay between the Colorado River and the Utah border. The strip was a vast, almost treeless sea of undulating land. Only a few people lived within its 12,000 square miles. Most, like Elmer Jackson, preferred to commute from the more sheltered surrounding communities. Thousands of cattle roamed the strip. There was not much else, unless coyotes, rattlesnakes, and desert iguana were counted.

It was this vast open space that was both the glory and the undoing of the American West. The vast spaces and small numbers of people attracted those activities that were shunned elsewhere. Scattered about the West was the technology of massive death. There were bombing ranges and atomic test sites, manufacturing plants for nerve gas and nuclear weapons, air bases and missile launching pads, and nuclear waste dumps and uranium waste piles.

The downwind region had extensive experience with both the beginning and the end of the nuclear cycle. Uranium prospecting and mining boomed from Tonopah to Fredonia and beyond during the 1950s, when the AEC encouraged greater domestic production in order to build up its stocks. Prospectors frequently got high readings on their radiation detection devices from the fallout that lay on the surface of the ground. If the topsoil was scraped away, the readings would drop dramatically. Not everybody knew this, and some worthless mining claims were sold to gullible buyers at outrageous prices.[10]

Fredonia, which means "free women" according to local lore, was founded on the northern border of the strip in the 1880s by polygamists who were escaping prosecution in Utah, where plural marriage was illegal. Vonda McKinney, who initiated the Allen suit, wrote a one-page history of Fredonia that was included in a county history. This was Fredonia in the 1950s:

> Today, although still a small community with a population of but five hundred, Fredonia boasts an elementary and an accredited high school, a beautiful chapel, several motels, cafes, service stations and stores, and a large milling operation as well as farming and ranching. Gateway to the Arizona Strip country, its possibilities of growth and improvement are great.[11]

From Fredonia the road, and the wind-driven clouds, passed on to the east.

First, the sheep died. The sheep were surrogates for humans. They should have been regarded as an early warning so that precautions could have been taken for the people. Instead, the AEC went to great lengths, including deception and fraud, to put the blame for the sheep deaths and injuries on anything but radioactive fallout. The sheep deaths and the subsequent coverup of the evidence of the possible cause served as precedents for the human experience. One of the people involved in the coverup was Warren E. Burger, who was then an assistant attorney general in the Department of Justice and who was to become Chief Justice of the United States Supreme Court. It was a shoddy moment in the nation's history.

The spring of 1953 was unusually dry in the west deserts, those warmer lands in Nevada where large numbers of sheep grazed during the winter months before being trailed back to ranches in the Cedar City area in the spring. The condition of the range was not good, but neither was it poor. The ranchers had experienced worse droughts. They brought in supplementary feed and water for the sheep as they began to drift back toward Cedar City for lambing and shearing.[12]

There was also an unusual amount of fallout that spring from the atomic tests at the nearby test site, particularly on the land north of the test site from Shots Nancy and Harry. The test organization made no attempt to determine the locations of these scattered groups of sheep and their herders prior to a test. A plan to tell livestock owners about radiation levels had been abandoned early in the Upshot-Knothole series.

Beginning in April and extending through May, the ranchers were first puzzled and then worried about the large sheep losses they were suffering. Of 11,710 sheep within an area up to 40 miles north of the test site and 160 miles east, 1,420 lambing ewes and 2,970 new lambs died. Other sheep displayed beta burn injuries similar to those found on cattle and horses around the Nevada site and at Trinity.[13]

Some sheepmen faced financial ruin. One such Cedar City area rancher was Douglas Clark, chairman of the commission that governed Iron County. Clark suffered some of the heaviest losses that spring. AEC officials and Stephen Brower, the county agricultural agent in Cedar City, visited his ranch. Among the AEC officials was Lieutenant Colonel John H. Rust, a veterinarian who was an expert on radiation and sheep. Rust later became a professor of radiology and pharmacology at the University of Chicago.

Clark asked the AEC men questions. There was a confrontation. Brower said, "Rather than answer his questions, the colonel verbally attacked him, calling him a dumb sheepman, stating that even if he gave him the answer he wouldn't be smart enough to understand it, etc. For 10 or 15 minutes the colonel harangued, belittled, and discredited Doug's intellectual ability to understand anything on this subject."

Later that afternoon the rancher died of a heart attack. An AEC official, Joe B. Sanders, noted for his files: "Mrs. Clark stated, to people who inquired about his condition, that he had been through many trying circumstances during the week. One of them could well have been a somewhat warm discussion between him and Rust." Clark may have been, in an indirect sense, the first human casualty of fallout from the Nevada site.[14]

The first outside veterinarians to examine the sheep were John I. Curtis, the state veterinarian, and F. H. Melvin, who was assigned to Utah by the Bureau of Animal Industry within the Department of Agriculture. In company with the puzzled local veterinarian, the two men from Salt Lake City examined a number of herds in the Cedar City area in late May. The ranchers believed that fallout was to blame, but the two Salt Lake veterinarians voiced no opinions at the scene. They returned to the city, and Curtis mentioned to Dr. George A. Spendlove, director of the Utah Department of Health, that radiation was a possible cause. Melvin said he had never seen anything like it before.[15]

A worried state health director then requested epidemic aid from the Public Health Service. Three veterinarians were promptly dispatched to southern Utah: Monroe A. Holmes, assigned to Utah from the PHS Communicable Disease Center; Arthur H. Wolff, veterinary radiologist at the Environmental Health Center; and William G. Hadlow, veterinary pathologist with the Rocky Mountain Laboratory of the PHS.[16] At Cedar City this party met two researchers recruited by the AEC: Major R. J. Veenstra, an army veterinarian attached to the Naval Radiological Defense Laboratory in San Francisco, and R. E. Thompsett, a veterinarian who had a private practice in Los Alamos and who was hired by the AEC on a contract basis. Thompsett was familiar with the Trinity cattle and had participated in the AEC investigation of Floyd Lamb's cattle the previous year. Joe Sanders, acting field manager of the AEC's Las Vegas office, accompanied them.

The two teams joined forces and toured the ranches in early June. They saw that most of the lambs had been born dead and in a stunted condition. Those that struggled to live gave up after about five or six days. Either they were too weak to nurse or the ewes had little or no milk. The ewes died either during lambing or a few days later. It was a pitiful sight.

Radiation and malnutrition emerged as early candidates for causes. A radiation detection device was passed over the throat of a sheep that had been slit open on the cement floor of the lambing shed. A rancher heard one of the AEC investigators remark, "This sheep is hotter than a $2 pistol." Brower, who accompanied the investigators, was told by Veenstra and Thompsett that the readings were quite high. Holmes noted that the Nevada range experienced one of the worst droughts that year. Although the winter was mild, the spring months were cold. There was not much new growth on the range.[17]

After a number of trips to the Cedar City area that June and numerous meetings among the various participants, Holmes wrote the main report for the seven government agencies that eventually became involved in the first investigation. Fallout and malnutrition were blamed for the deaths. The report did not decide between the two possibilities. It stated, "There are too many variable factors, and in view of the fact that the investigations were conducted so late in the stage of this affliction, much of the data is by hearsay and not by observation."[18] Separate reports filed by individual participants in the study, however, mentioned radiation more prominently. Wolff noted the high radiation levels in the sheep's thyroids. To Veenstra, "Radiation was at least a contributing factor to the loss of these animals." Thompsett concluded, "Again, with the sheep losses I am of the opinion that the Atomic Energy Commission has contributed to great losses." He suggested that livestock be grazed farther from the test site.[19] These reports, and others on the matter, were immediately classified. Local officials, like Brower, who had been promised copies never received them. Behind this wall of silence, the AEC began gradually to shift the blame from radiation to malnutrition.

Back in Washington there was quite a bit of hedging. Gordon Dunning reported to the AEC commissioners in June that horses and sheep near the test site had been determined to have beta burns on their backs and around their nostrils, while the death of cattle was attributed to malnutrition. The sheep deaths were being investigated. One month later Dr. Bugher, Dunning's boss, told the commissioners that the death of the sheep could "not have been caused by radiation; however, since the animals might have suffered some radiation injury it is possible that this was a contributing factor in their deaths." That was a rather murky statement, but the cause of the sheep deaths was to become even more clouded. In mid-July, Bugher reported that the horses did suffer beta burns but that the death of the cattle had not been caused by fallout. The sheep may have died from ingestion of toxic plants, said Bugher, although "they had apparently also suffered some radiation injury." [20]

In early August the AEC's Division of Information Services circulated a draft press release to the general manager and the five commissioners. The release was to be finalized after the division received comments back and the sheep investigation was completed. There was some talk about also obtaining the input of the "psychological strategy authorities." The draft release stated:

> Livestock grazing in the immediate vicinity of the proving ground have suffered some nominal injury from radioactive fallout. (Here supply the facts as are finally determined.) The livestock herds and flocks affected were in close-in areas where their grazing was a calculated risk on the part of the owners. Nevertheless the AEC has moved to compensate. (Here what the facts turn out to be.) [21]

The AEC did not in fact compensate the sheep owners. Such an action would have been an admission of guilt and incompetence, which the agency and its bureaucratic hierarchy ruled out for the sake of national security and the furtherance of personal careers. There would be no justice or compassion for animals or humans.

The search was on for another cause. One way to get different results was to employ different investigators—namely, those investigators who were more accountable to the parent organization. The Division of Biology and Medicine took over direction of the investigation on orders from General Fields, director of the powerful Division of Military Applications. Dr. Paul B. Pearson, chief of the medical division's Biology Branch, became Bugher's man in the field. The technical input now came mostly from Lieutenant Colonel Bernard F. Trum, an army veterinarian assigned to the

AEC's Agricultural Research Program at Oak Ridge, Tennessee. Trum was assisted by John Rust from the same organization. Thompsett, Veenstra, and the Public Health Service veterinarians were frozen out of further participation in the field investigation, which now centered on malnutrition as the cause.[22]

The new group toured southern Utah, and Pearson reported to his superior that they had noted the dry range conditions and lack of feed. He said there was very little evidence that the livestock deaths had been caused by radiation but that there was evidence pointing to malnutrition. The AEC was now hearing what it wanted to hear.[23]

While in Cedar City, Pearson visited Brower in his county office. Brower's account of the conversation has been given publicly four times since early 1979. Three of the times Brower was under oath. Pearson later said in a sworn deposition that he had no recollection of the conversation, to which Brower replied at the Allen trial, "I'm not surprised."[24] It was one man's word against another's. Pearson, who was a native of Utah, attended Brigham Young University. He left the AEC for the University of Arizona, where he taught in the Department of Biology and Medicine. Since his days as the Iron County agricultural agent in the mid-fifties, Brower had obtained a doctoral degree from Cornell University, had served the Mormon church in a number of positions, and had been a professor in the Graduate School of Management at BYU. At the trial, he was the very picture of Mormon rectitude with his three-piece gray suit and silver hair. Brower testified about the meeting with Pearson:

> He indicated there was no possibility that the sheepmen would be able to get reimbursement for damages from the AEC. He also indicated that the AEC could not afford to have that precedent set, and described some allegations on the part of people in the Midwest who were trying to get reimbursement for alleged damages from radiation.

Brower continued:

> It was a lengthy discussion, your Honor, dealing with this general subject matter. I was trying to establish both what the

AEC would and could do—what procedures we might follow
to elicit help from them. At one point he indicated he was a
Utahan. He said he felt badly about the problems of the sheepmen and offered, if the ranchers would cooperate, to provide
$50,000 for a range nutrition study. They subsequently did
provide $25,000 for such a study.[25]

Brower had told the same story at the April 1979 congressional hearing, where there was this exchange with Senator Edward Kennedy:

> KENNEDY: You mean they were told to forego their claim
> because if they settled with one, they'd have to settle with
> others?
>
> BROWER: Exactly.
>
> KENNEDY: And if they didn't cause any problems, they'd
> get a range nutrition research program?
>
> KENNEDY (not allowing Brower to complete his answer):
> Does that make any sense from your point of view . . .
>
> BROWER: Doesn't make any . . .
>
> KENNEDY: . . . in terms of equity or fairness or responsi-
> bility?

It did not make sense twenty-five years later to Kennedy, but at the
time, Brower and the ranchers were being worn down by the constant assertions of AEC investigators who trooped through the area
proclaiming that radiation was not the cause of the problem. Also,
they did not have access to the classified documents that surfaced
later.[26]

In early August, all those who participated in the sheep investigation, including the first set of experts, met in Salt Lake City to review
the evidence. From the start there was disagreement among the
twenty or so participants, one of whom was the omnipresent Gordon
Dunning. They were asked by Pearson, who conducted the meeting,
not to make their discussions public. Veenstra and Thompsett held
to their previous position that radiation was at least a contributing
cause. The others tried to change their minds. Dunning and Trum
were lined up against Veenstra and Thompsett. Pearson's report on

the meeting mentioned no dissent, and it inferred that all had agreed that radiation was not the cause.[27]

That was the basic message that Pearson took to Cedar City for a meeting with the ranchers on August 9, but he needed something to blame for the deaths. The scientist first tried out the idea of an unknown livestock disease and then tried malnutrition, neither of which the ranchers bought. The radiation that was measured, it was explained, was too little to kill. The investigation was continuing. No conclusions had been reached yet, the ranchers were told.[28]

Meanwhile, the AEC sought to give scientific credence to its predetermined policy. Experiments on sheep were conducted at the Los Alamos Scientific Laboratory, and it was supposedly demonstrated that it took exceedingly high doses—higher than any AEC estimates of what the Utah sheep had received—to inflict beta burns on sheep. Similar results were obtained at the AEC's Hanford laboratory. Of course, the dose estimates may have been too low or the experiments conducted in a faulty manner, as suggested at the time by John J. Finn, a government lawyer, and later by Harold Knapp, a former AEC scientist.[29]

With this new evidence in hand, the AEC sought unanimity. A meeting was held at Los Alamos on October 27 to evaluate these studies. Most of the veterinarians who had previously participated in the investigation were present. Veenstra was the only dissenting investigator who was not present. Gordon Dunning attended as a representative of the Division of Biology and Medicine from AEC headquarters.

At the morning session in the Health Research Center, colored slides of the experimental sheep were shown along with pictures of the biopsy sections of sheep that had been slaughtered in southern Utah. The group then inspected the irradiated sheep. Back at the meeting room that afternoon a discussion was held, and Dunning, who served as secretary for the group, sought to summarize it in a one-page statement, which concluded: "All of these data present a preponderance of evidence to support the conclusion that the lesions were not produced by radioactive fallout."[30]

The participants signed their names to the typed statement, but not all of them agreed with it. Holmes, noting disagreement among the group with Dunning's conclusions, later reported that "we were asked to sign this attesting to our attendance, which I did." Wolff noted in his report on the meeting, "Dr. Dunning indicated that it was imperative that he prepare a statement for Commissioner Zuckert of the AEC pertaining to the Utah sheep situation. This statement, claimed Dunning, is required before Commissioner Zuckert will open up the 'purse strings' for future continental weapons tests. Accordingly, a statement was agreed upon by the group heretofore listed."[31]

Back in Washington, Dunning said of the seven signatures, "I doubt if we will ever obtain more positive conclusions from this committee than are contained in the attached statements." He urged that a joint press release be issued to announce their agreement on the cause of the sheep deaths.[32]

The following day the Division of Biology and Medicine informed the AEC commissioners that fallout was not to blame for the sheep deaths. "The preponderance of evidence indicates that activities of the Commission at the NPG were not responsible for, nor did this contribute to the heavy losses of sheep that occurred in the Cedar City area this spring." Nevertheless, the report noted, there was not complete agreement on this conclusion, so further studies and investigations would be undertaken.[33]

It was doubtful that the commissioners knew how far from unanimity the experts were. The bureaucrats acted as buffers. In reply to a letter from Pearson, Lieutenant Colonel Veenstra of the Naval Radiological Defense Laboratory challenged Dunning's calculations, stating that they were "valid but had not gone far enough." Holmes, the PHS veterinarian, poked holes in Pearson's report. Thompsett, the Los Alamos vet who had had experience with previous radiological burn cases, thought the AEC was "really in trouble." It was significant that all three men were outside the ranks of the AEC.[34]

The AEC issued a press release on January 13, 1954, stating that radiation did not cause the sheep deaths, but no cause was cited in the release. AEC personnel unofficially suggested to the press that poor range conditions and poisonous plants were the cause, and these factors were dutifully reported. Actually, a Department of Agriculture study had previously ruled out toxic plants.[35] After mentioning poor range conditions and toxic plants as possible causes, a *New York Times* story noted:

> The episode presented the AEC with one of the most ticklish situations since it began its continental tests. The agency was certain before it started the tests that they would involve no hazard to anybody or anything off the test reservation. Something like $10,000,000 has been invested in permanent facilities at the test site, and the commission's development program is geared to its operation. Yet, if a thousand sheep had been killed by radiation, the inescapable inference was that it might have been a thousand human beings.[36]

All that seemingly remained was for AEC officials to convince the sheepmen of their conclusions. Another meeting was held in Cedar City that month. It was inconclusive. The ranchers were told that the range nutrition study might come up with some definite answers. The AEC officials at the meeting attempted to discourage the sheepmen from filing claims or suits.[37]

With the resumption of tests at the Nevada site in early 1955, five suits were filed by ranchers in the federal district court in Salt Lake City asking the AEC to pay a total of $177,000 for the sheep that were thought to have been killed or maimed by radioactive fallout two years earlier. The AEC was in serious trouble, and the second phase of the cover-up now began in earnest. There were those experts who did not agree with the AEC's position on the cause of the sheep deaths. Potentially damaging evidence and testimony had to be suppressed or altered. The AEC staff and the Civil Division of the Justice Department, which was headed by Warren Burger, cooperated to accomplish these tasks.

The first move the AEC made was to take soundings among the offsite population. Major Grant Kuhn, a veterinarian attached to the AEC, left the test site by the back, or north, gate in early March and toured the ranches from Tonopah to Alamo to Cedar City, attempting to assess the situation and reassure the stock owners of the safety of past and present tests. One rancher, Kuhn reported, "tested me on where I stood on government redress of ranchers. I explained to him that I was only interested in learning the truth about the effects of fallout on livestock through the facts." Another rancher, after talking with Kuhn, said that he was not going to join the suit. At the Fallini ranch Kuhn inspected the cattle. A fallout cloud had passed over the ranch eight days earlier, and there were gray splotches on the backs of the cattle. After taking some readings and explaining the effects of radiation on livestock, the major reported, "They were satisfied that the amount they had experienced was of no consequence."[38]

On March 31, nine days after completing his initial survey, Kuhn departed on another tour of the area with Lieutenant Colonel Trum, who was the technical advisor to the government attorneys. Trum was scouting out evidence, seeking potential witnesses, and interviewing those who were familiar with the sheep problem. He visited one of the plaintiffs, who told the army veterinarian that he had many deformed lambs and was losing many new sheep. At Alamo, Trum was told by local residents, whom he termed "definitely not hysterical," that they did not trust AEC officials and believed that they had frequently been lied to.[39]

Having surveyed the attitudes of the plaintiffs and other offsite residents, Trum then went to work on Veenstra and Thompsett, who still held opinions that differed from the official AEC determination of the cause of the sheep deaths. Their opinions needed to be changed before they gave a deposition or appeared at the trial so that the government could present a united front and not inadvertently serve the interests of the sheepmen. The pressure on the dissenters was increased.

Trum wrote Veenstra a "Dear Bob" letter, stating, "I got the distinct feeling that you felt that there was a chance that radiation could have caused the death of some of the sheep at least." He cited additional sheep experiments that had been conducted at the AEC's Oak Ridge facility. The results were the same as for the experiments conducted at Hanford and Los Alamos. "Because you didn't have this information available when you were asked to make your statement," wrote Trum, "I've been wondering if you might not have changed your mind about these things? If you haven't changed your mind, I'd like to know what you are basing your opinion on, for I shouldn't like to go into this thing [the trial] divided within our own [Army Veterinary] Corps if we can avoid it."[40]

Without waiting for an answer, Trum visited Veenstra in San Francisco on April 1. Veenstra outlined his skepticism about the various experiments conducted at the AEC facilities. The accumulated effects of the total fission products were not adequately considered, he said, and the Hanford experiments were conducted on healthy sheep, not sheep that had been out on the range all winter and may have been in a weakened condition. Trum presented additional evidence in an attempt to counter Veenstra's reservations, and Veenstra promised to reevaluate his position. He was being worn down.[41]

Five days later Veenstra wrote a letter that was typed but never sent to Trum. His position remained the same. The letter stated, "Radiation could have contributed to the death of the animals." He pointed out that the Los Alamos and Hanford tests were faulty. On the same day, Veenstra made a tactical retreat. He sent a handwritten note to Trum stating, "If called upon for a statement, I will just repeat what we found and say we felt it a possibility that should be pointed out for consideration."[42]

Trum then went to work on Thompsett. The veterinarian was in a vulnerable position. He had already lost his consulting contract with the AEC, and he depended on his contract with the AEC to operate his small-animal hospital in Los Alamos. Trum sent Thompsett a "model letter" on May 9 and suggested, "If the letter is not exactly to your liking or not in your style, you make the changes as you see fit. I believe this fits the facts as we know them today." The veterinarian signed the letter after making some minor changes in the text. The letter placed him in accord with the position of the AEC as expressed in the 1954 press release. Thompsett told Brower at the time that he had been forced by his superiors to reverse his findings.[43]

Trum, who later went on to direct a primate research laboratory at Harvard University, was confident that he now had the testimony wrapped up in favor of the AEC. He made his report three days later to United States Attorney Llewellyn O. Thomas in Salt Lake City.

Of Veenstra's and Thompsett's testimony, he said, "They will definitely state that they do not have evidence that radiation injury *was* either the cause or contributed to the deaths of the sheep. They will disqualify themselves and leave to others the problem of demonstrating that ionizing radiation *did not cause* demonstrable injury to the sheep." Trum concluded: "In recapitulation I believe that the test group can effectively demonstrate that no hazard was intended or in fact no hazard existed in the area in which the sheep were grazing." With Thompsett's reversal and Veenstra and Holmes agreeing to disqualify themselves as experts, Trum gave the following breakdown on the remaining expert witnesses: Wolff "takes an 'I don't know' stand and Rust and Trum state 'it could *not* have been radiation.'"[44]

A Department of Justice attorney, John J. Finn, left Washington on May 23 to visit many of the same potential witnesses in order to complete and put into a form that could be used in the subsequent trial the arrangements that Trum had made. Finn was the chief trial attorney in the sheep case. He was in the Torts Section, which was headed by Bonnell Phillips. Phillips's activities were directed by Burger, who was in charge of the Civil Division.

Burger became chief justice of the Supreme Court in 1969, and the sheep case eventually wound up before the highest court in the land. Burger disqualified himself from ruling on the appeal. He gave no reason, but there was persuasive evidence that Burger was well aware of the machinations going on within the Civil Division in regard to the sheep case. If he was not, then he was badly remiss in his duties. A federal judge would later rule that what transpired within the Department of Justice in the mid-1950s was a fraud committed upon the court.[45]

Burger's name appeared at the end of a letter dated June 20, 1955. The letter was sent to the AEC's general counsel. It was an account of Finn's trip. Finn, accompanied by three AEC officials, had visited Thompsett on May 26 in his Los Alamos animal clinic. It must have been an impressive, and persuasive, entourage. The letter stated:

> In the conversation with him he advised that he had amended
> his findings and opinion previously enunciated and reported

to the Atomic Energy Commission and to the sheepmen personally. He states that his present position is taken because of additional data of a scientific nature furnished him by Doctors Trum and Lushbaugh, and perhaps somewhat influenced by other "scientific" evidence furnished him. At the conclusion of the conversation, Dr. Thompsett furnished Mr. King with a letter setting forth his present opinions relative to any connection between radiation and the lesions noted previously by him.

This was a reference to the "model letter." Clarence C. Lushbaugh, a radiologist with the Health Division at Los Alamos, conducted the sheep experiments at the laboratory. Assistant General Counsel Chalmers L. King of the AEC was a personal friend of Thompsett and had a number of conversations with the veterinarian on the matter.

The government attorney then visited Veenstra at his San Francisco laboratory on June 6, according to Burger's letter. Finn was accompanied by Leo Bustad, who had conducted the experiments on the sheep at Hanford. The letter stated:

It will be recalled that Dr. Veenstra concurred in the original opinion of Dr. Thompsett and he states he has had no evidence presented to him to this date to indicate that he should form a different conclusion now. In fact, he furnished a copy of a letter which he had written but never sent to Dr. Trum, setting out in full his present position, which obviously is at variance with the conclusions of Dr. Trum and of others, including Dr. Thompsett's present views. After some discussion with Col. Veenstra and Mr. Charles Sondhaus, a radiation physicist, they agreed that they, because of additional experiments which had been conducted in regard to sheep, which were not known or made available to them, would be disqualified as experts in these cases. They would have to state that they had not enough data to be qualified to give an opinion as to the cause of the lesions which were observed at the time the sheep were shown at Cedar City, Utah. Of course, astute plaintiffs' counsel can make capital if Col. Veenstra is interrogated.

Finn visited others on his trip, including Kermit Larson at UCLA. Burger reported, via Finn, that Larson "will be valuable" at the trial. Finn met with Melvin, one of the first veterinarians to inspect the sheep, and reported that he would also be helpful. He also thought that Stephen Brower and Colonel Rust would make good witnesses. Finn then returned to Washington and wrote his report.

On the same day that the report on Finn's trip was sent to the AEC, Burger sent a second letter to the AEC's general counsel requesting that an AEC investigator, James B. Malone, gather information that could be used in the trial. The government lawyers needed eighteen items of information, including tax returns and statements to banks by the ranchers, so that the number of sheep that they owned could be ascertained. The investigator was also asked to obtain all the information he could on the drought. Two government versions of the locations where fallout occurred during the spring of 1953 existed and needed to be reconciled and an official position established by Malone.[46]

Before the start of the trial, the attorney for the sheepmen, Dan S. Bushnell, submitted a set of written questions, called interrogatories, to the government attorneys. He asked, "Did anyone involved in the investigation disagree with the reports?" The government lawyers labored mightily with the answers, and there was a lot of cross-checking between the AEC and them before they were sent off. The answer was: "We are not aware of anyone who is involved in the Commission's investigation of the alleged sheep loss who now disagrees with the report issued by the Atomic Energy Commission." The key word, of course, was *now*, and the switch from the past to the present tense was deceptive.[47]

The trial got underway in September 1956 and lasted fourteen days. Finn and the other government attorneys were aggressive and spared no resources. Bushnell, a young Salt Lake City attorney, represented the plaintiffs. Judge A. Sherman Christensen listened to the evidence without a jury, since the suit was filed under the Federal Tort Claims Act. Thompsett did not testify at the trial, and the other expert witnesses testified as predicted. The radioactive fallout and the sheep deaths were a coincidence, said Finn.[48]

Judge Christensen handed down his decision at the close of the trial. He ruled against the plaintiff, David C. Bulloch, whose suit was used as a test case. The sanitized testimony was persuasive. The judge noted, "Of the three professional men who originally suggested radiation damage, two, upon further consideration, questioned their original diagnosis. None of them claimed to be particularly qualified in the field of radiation. On the other hand, some of the best informed experts in the country expressed considered and convincing judgment that radiation damage could not possibly have been a cause or a contributing cause." Christensen said the government had been negligent in not warning the sheepmen of anticipated fallout.[49]

The case was not appealed. A member of the staff of the Joint Committee on Atomic Energy noted in the margin of a letter from an AEC attorney enclosing copies of the opinion: "Very interesting. Does this raise questions to AEC advantage in having many 'hired' experts available?"[50] It certainly did.

The federal government thought it had successfully contained any damage that might result from the suit. The cover-up rose to the surface, however, a quarter century later to haunt the present-day successors to those who had previously handled the matter. A wealth of new information that had once been suppressed was made public, and another viewpoint emerged.

At the 1979 congressional hearing in Salt Lake City, F. Peter Libassi, the general counsel of the Department of Health, Education, and Welfare, noted that radiation dose levels to the thyroids of sheep were nearly one thousand times the permissible dose for humans. Libassi, who also headed an interagency task force on radiation, testified: "As we can read from those memos, it was the Public Health Service officials who believed that the conclusion was not warranted by the facts at that time." (But those officials had remained publicly mute at the time.) Dr. Donald S. Fredrickson, director of the National Institutes of Health, stated, "I think, Senator Kennedy, on the basis of the documents that I have had the opportunity to see, it would have been extremely difficult, probably impossible, to conclude that radiation did not at least contribute to the cause of the death of the sheep."[51] The congressional committee's report, *The Forgotten Guinea Pigs*, stated that evidence concerning radiation as a cause of the sheep deaths and injuries "was not only disregarded but actively suppressed," and that the government had fabricated reports.[52]

Six of the original plaintiffs filed suit in the federal district court in Salt Lake City in February 1981, alleging that a fraud had been committed upon the court and asking that the decision in the 1956 sheep trial be set aside and a new trial held. After paying off debts from losses incurred in 1953, only two of the original seven families who had filed suit in 1955 were still in the sheep business.[53]

Twenty-five years after handing down his first opinion, Judge Christensen, a respected jurist, was the senior judge of the court. Bushnell, now in his late fifties, was a partner in a successful Salt Lake law firm. He was one of the unsuccessful candidates for the federal judgeship that went to Bruce Jenkins, the judge in the Allen case. The government was represented by Henry Gill, who also headed the team of lawyers in the Allen trial. Before filing suit, Bushnell had offered to settle the case with the federal government for $3 million. The government refused the offer. Judge Christensen heard testimony for four days during May 1982 and handed down his startling decision in August, one month before the beginning of the Allen trial. It was a devastating indictment of the government's behavior.

Running between the lines of Judge Christensen's opinion was a thread of barely controlled anger. There was also the sense of innocence lost. The fifties was a time of greater trust in the government. It was a time, before Watergate, the judge said, when he had "a somewhat pristine view" of the integrity of government officials. Moral values had changed. "Nonetheless, I have concluded that by whatever standard or as to whatever period the circumstances found here are to be judged, they clearly and convincingly demonstrate a species of fraud upon the court for which a remedy must be granted at even this late date," wrote Christensen. This was no ordinary case, he continued, since it involved "a transcendent chapter of world history" and "a mysterious and awesome device as to which the AEC and those associated with it enjoyed a virtual monopoly of knowledge." He wrote:

> In such a setting it appears by clear and convincing evidence, much of it documented, that representations made as the result of the conduct of government agents acting in the course of their employment were intentionally false or deceptive; that improper but successful attempts to pressure witnesses not to testify as to their real opinions, or to unduly discount their qualifications and opinions were applied; that a vital report was intentionally withheld and information in another report was presented in such a manner as to be deceitful, misleading, or only half true; that interrogatories were deceptively answered; that there was deliberate concealment of significant facts with reference to the possible effects of radiation upon plaintiffs' sheep; and that by these convoluted actions and in related ways the processes of the court were manipulated to the improper and unacceptable advantage of the defendant at the trial.

Specifically mentioned within this context were Trum, Rust, and Pearson of the AEC.

Of the actions of the government attorneys, Judge Christensen wrote that he found "that one or more of defendant's attorneys knowingly participated in a program for the concealment from the Court of facts which he or they knew or in good conscience should have known the Court was entitled to have placed before it in order to properly rule upon the determinative issues of the case." Specifically mentioned in this context were Finn and King of the Justice Department and the AEC, respectively. Judge Christensen concluded that a new trial should be held.[54]

A decision of fraud being committed upon the federal court, especially by the federal government, was extremely rare. The Justice Department later admitted in its appeal that "the district court's decision was a sweeping indictment of the government's conduct in Bulloch I." It found some solace in the fact that "only four supposedly fraudulent actions actually were specified."[55]

Neither Burger nor Phillips were mentioned in the opinion. Both

Christensen, a member of the federal judiciary that Burger headed, and Gill, a government attorney, leaned over backwards not to implicate the chief justice and the head of the Torts Section during a brief exchange at the trial. In effect, Judge Christensen was covering up a cover-up, and everyone—including Bushnell, who did not object —seemed willing to go along with the conspiracy of the bar.

The judge first inspected the Burger letters in his chambers and then said in open court that he objected to the way that Gill had listed them as exhibits. "Mr. Gill," said Judge Christensen, "I'm sure that in that listing you didn't intend to indicate that even the head of the Torts Section, and certainly not the Assistant Attorney General of the United States, was in there participating or being involved in writing about these tactics and all that sort of thing?"

Gill replied, "Not at all, your honor."[56]

But they clearly were involved.

It took longer for the people to die, and then great sorrow followed. The personal stories of the plaintiffs were exceedingly painful to hear. The atmosphere in the austere courtroom was heavy with the searing emotion of long, painful cancer deaths and disablements being recalled in excruciating detail. The intimacies of lives together and the wrenching separation of death were revealed to public scrutiny by the penetrating questions of lawyers. The witnesses, most of whom had never before been in a courtroom, were nervous. The relatives of the dead cancer victims sometimes broke down on the stand. At those times, the spectators, too, were visibly touched. Judge Jenkins said at one point, "I know it is hard. It is hard on me."[57] It was hard on everyone.

Twenty-four test cases, which were supposedly representative of the nearly 1,200 plaintiffs, were chosen by the claimants' lawyers to be heard at the trial. Using "bellwether" cases made the process more manageable. It was hoped that these representative cases, after the initial decision and review by higher courts, would provide a pattern for the remainder.

The first group of plaintiffs consisted of twelve males and twelve females, of whom four were youngsters who had died between the ages of twelve and fourteen. Nineteen were dead at the time of the trial, and five of the plaintiffs were living. Nineteen lived in Utah during the period of atmospheric testing, three in Nevada, and two in Arizona. Their homes and workplaces were either east or north-east of the test site. There were eight leukemia deaths, one case of Hodgkin's disease, a lymphoma, and cancers of the lung, stom-ach, brain, skin, uterus, ovaries, breast, pancreas, kidney, bladder, colon, prostate, and thyroid. Most of these cancers can be induced by radiation or other causes. All of the cancers involved great suffer-ing.

It would be hard to imagine a more homogeneous group, given the randomness of cancer induction. Most were practicing Mormons and reflected the everyday virtues of that religion. They did not smoke and did not drink liquor, coffee, or tea, all of which can increase the risk of cancer. Families were large and exceedingly close. The wage earners were industrious, but incomes were mod-est. Usually the husband and wife both worked, and invariably the woman worked when widowed. There was a great deal of constancy among these people. Only a few of the men and women who were widowed ever remarried, and those who did tended to have un-happy second marriages. Spiritually, they remained joined to their first mates.

Virtue did not necessarily bring rewards, however, since its prac-tice contained the seeds of destruction for some. These people were thrifty and self-sufficient. Fresh vegetables and fruit were grown in backyard gardens and orchards. A family usually owned a milk cow or purchased raw milk from a nearby dairy. Both adults and children drank raw milk in large quantities. Beef, lamb, pork, and poultry were slaughtered locally. In this manner, radiation can make its way through the food chain and into the human body.

Most were aware of the atomic flashes, blasts, and clouds; a few were not. Few could recall being in visible fallout, but that did not mean that they had not been exposed to it, since fallout could also be invisible. All said that they had never received any warnings from the government about the health effects of fallout. None had defi-nitely associated fallout with cancer until the late 1970s. Almost all of these people made the long drive to a major population center—Salt Lake City for most—where they underwent diagnosis, surgery, or painful radiation and chemotherapy treatments. The drive back was a long journey laden with sorrow and tears. Few survived many such treks, and their families never forgot them.

The majority of the plaintiffs were not particularly interested in the money they might gain from a successful lawsuit. They were more interested in protecting future generations by establishing the harm as a fact. These people had a great sense of obligation to others of their kind. A few did seek money. The widow of one cancer victim wrote her lawyer, "I am not like many of your clients who don't care about the money. I want all I can get plus medical benefits and other things. That's the reason for a lawsuit."

There was unanimity on one goal. They were looking for a gesture, an admission from their government. None was forthcoming. In fact, through the adversary process of cross-examination, the government exhibited a hostility toward the witnesses. This did not seem appropriate, or just, or compassionate.

These people were known as the "at risk" population to the AEC, "plaintiffs" to the court, and "victims" to history. They were certainly victims of cancer, but it remained to be seen if they were all victims of fallout. They are best encountered in the order of their deaths or, if still living at the time of the trial, in the order of their diagnosis. Because leukemia had the shortest latency period of any of the cancers, those victims died first. For each victim a bell tolled.

Karlene Hafen spent her entire short life in St. George. She died at the age of fourteen on November 17, 1956—a little more than three years after Shot Harry. Leukemia has a minimum one- to two-year latency period.

The atomic tests were exciting events for the Hafen family. They were history in the making. The three Hafen children, of whom Karlene was the eldest, were awakened in the predawn hours to watch the flashes over the black hill to the west. Sometimes what appeared to be dust later floated over the town. On some occasions there were warnings to keep the children inside, but no reason for such warnings was ever given.

Karlene was a beautiful, sweet child. She liked to sing, was a straight-A student, and was a candidate for Harvest Queen at the dance in late November. She did not live to attend the dance.

Four months earlier Karlene had been a flower girl at a baby's funeral. She fainted at the cemetery. A local doctor, after conducting some tests, advised LeOra Hafen to take her daughter to Salt Lake City, where she was diagnosed as having acute myelogenous leukemia. There were two more trips to Salt Lake.

On November 11, Karlene said, "Mom, call Dr. Cowan. I am going to die."

"Oh, you are not either," said her mother.

"Yes I am," insisted Karlene. "I am just full of angels."

Six days later she died. Her father, a rancher, died of lung cancer in 1965. Other family members also had various types of cancer.[58]

Sheldon Nisson lived in Washington, a small hamlet just east of St. George. The young boy was particularly close to his father. Sheldon drove the tractor on the family farm at the age of ten, after having been shown how by his dad, and the two went hunting together. Sheldon was smart in school. His mother thought he might grow up to be an astronaut, since he liked to build rockets and airplanes. He was a leader of his church group, and when he painted, Helen Nisson said, "It just looked like the Lord had ahold of his little hand."

At Easter in 1959 the Nisson family, which included three other children, went on a picnic. Sheldon did not have his usual high spirits. His mother thought he had the Asian flu, which was going around then, but a local doctor referred Sheldon to a Salt Lake City cancer specialist, who diagnosed his illness as acute myelogenous leukemia.

The Nissons were told that their son did not have long to live. "He was so thin and pale, and his little eyes looked so pitiful; but he had little curly hair and it was so cute, and he had so much to live for," his mother lamented. The boy was not told that he had the dreaded disease, which was almost always fatal in those days.

But Sheldon knew he was going to die. His mother told him to hurry up and get better so that he could go hunting with his dad.

"I don't think there will be much hunting for me," said Sheldon. He died in his father's arms on July 6, 1959, at the age of thirteen.

The cemetery in which young Nisson was buried was on a bare red-dirt hill. Red—dirty red, to be precise—was the color of the fallout clouds that Mrs. Nisson saw float over Washington. Operation Upshot-Knothole had concluded only two weeks before Sheldon's seventh birthday. There was no danger, said the AEC. "We believed the government," said Mrs. Nisson. "The government was the most marvelous thing. We thought they wouldn't do anything wrong."

Darrell Nisson, who operated a rock quarry along with the farm, did a little prospecting on the side. He combed the hills of Nevada and Utah looking for uranium. "Hell, everybody was going to be millionaires around here for a couple of years," said Sheldon's father. As he traversed the countryside with his Geiger counter, he noticed that some spots were hotter than others from the fallout.

Sheldon's death hit his father hard. He did not want to discuss it. When Sheldon's name was mentioned, his father shuddered and remained silent. His health declined for no obvious medical reasons. At the time of the trial, Nisson was in a rest home.[59]

Peggy Orton bumped her foot on a cement step and broke a bone in it. The doctor noticed Peggy's lassitude and paleness and conducted some tests. The diagnosis was acute lymphoblastic leukemia. Six months later, on May 29, 1960, the fourteen-year-old girl died. Like other families, the Ortons, who lived north of Cedar City in Parowan, got up early to watch the tests.

A teacher of Peggy's, Jean Hendrickson, took her son up to a high point just outside of town to watch the blasts and resulting clouds. She said, "It was history. This was for the United States Government." Mrs. Hendrickson never received any warnings at school to keep the children inside, nor was she aware that the clouds were hazardous. "We trusted our government," said the teacher.

After the diagnosis, Peggy returned to school. At times she had to be taken home by her teacher because she felt nauseous or was vomiting. The young girl was not told that she had leukemia. "I especially remember Peggy during a Christmas program," said Mrs.

Hendrickson. "She had beautiful red hair, most outstanding, and the whole group of little girls was standing on the stage in the Christmas program, and they were all holding candles, and Peggy held this candle and her face was quite puffy and very white. It was just something that I never forgot."

After Peggy's death, her father began drinking. He spent hours at the cemetery. Although the family had contracted for perpetual care, he planted grass on the grave site and watered it.[60]

Lenn McKinney was the fourth to die. The letter from his elderly mother to his wife began the lengthy process that resulted in the legal challenge to the government. The McKinneys were one of the most active couples in Fredonia. Lenn was on the town council and the school board. He quit working for the lumber company and started his own trucking business, hauling lumber out of town and cottonseed meal and salt into town for the ranchers, and building supplies and coal for the town dwellers. McKinney owned as many as three trucks. Mrs. McKinney was president of the Parent-Teacher Association, the Boosters Club, and the Women's Club. She was also justice of the peace and the unofficial coroner for the area, the county seat being 200 miles distant in Flagstaff. In addition, she ran a service station and motel, a cafe, and then a grocery store. The couple had three children.

To their friends, the McKinneys seemed like extremely busy people. They worked hard, frequently for sixteen to eighteen hours a day, and they began to prosper. As the money came in, they invested in property—a house in Phoenix, two in Holbrook, their home in Fredonia plus two rental houses, two commercial buildings, a small subdivision, and ten acres of farmland. The couple was also happy. "I always said that I had one man out of ten thousand," said Mrs. McKinney.

Lenn McKinney complained of being unusually tired in the spring of 1961. When driving a truck back from Phoenix, his nose bled uncontrollably. He was diagnosed as having acute myelogenous leukemia in June, and he died on May 14, 1962. McKinney neither smoked nor drank. There was no family history of cancer. Before he died at the age of fifty, McKinney said he thought he had caught the disease from his daughter's best friend, Odessa Burch. At the time it was thought that leukemia might be contagious.[61]

The picture of *Arthur Bruhn* in the 1953 yearbook of Dixie Junior College showed an earnest, serious face with a slightly receding hairline and rimless glasses. There were twenty-nine instructors on the college's staff, and Bruhn, who was chairman of the Division of Biological Sciences, taught a wide variety of science courses. That spring Bruhn was in his midforties, the father of four children and an ambitious educator who, with the help of a master's degree, had risen from teaching elementary school to his present position. The next year he would be named president of the college.

It was during the Upshot-Knothole series of tests that Bruhn, in company with the geology class he took to witness the shots, was exposed to a fair amount of radiation. Besides having a great deal of scientific curiosity, Bruhn was an avid photographer and took pictures of some of the shots. On the day of Shot Harry, Bruhn called his wife from the college and told her to keep their children indoors. She had not been listening to the radio in their St. George home.

Bruhn was dedicated to the college. A popular teacher, he was known to his students as "A. B." or "high pockets" because he wore his pants high and short. As an administrator, he was somewhat dogmatic and stepped on a few toes. Bruhn was under a certain amount of stress. There were difficulties with the faculty, and he was being pressured to complete work on his doctor's degree so that he could retain the college presidency. He felt tired.

The educator was diagnosed as having acute lymphoblastic leukemia in late 1963. While hospitalized in Salt Lake City for treatments, his room was inundated with former students who had come to visit him. When Bruhn returned to St. George, he attended a basketball game at the college. He walked into the gymnasium late. The game stopped and all rose in a silent tribute. Bruhn died in the summer of 1964. Like McKinney, he had been struck down in the prime of his life.[62]

The last youngster to die was *Sybil Deseret Johnson*, the middle child of five. She was known as the peacemaker among her brothers and sisters. Her long hair, shading from red to gold, was quite noticeable. Sybil's parents taught music, and she played the cello in the high school and college orchestras in Cedar City. Perhaps some-day she would have been an accomplished musician. Instead, she died on May 15, 1965, at the age of twelve of acute lymphoblastic leukemia.

Her mother thought that she was the one who should have been taken. Cancer ran in the family on the mother's side. The Johnsons remembered seeing the fallout clouds. Sybil may have been exposed *in utero* or as a baby during the Upshot-Knothole series.[63]

Lavier C. Tait of Fredonia died on September 13, 1965, at the age of thirty-seven. A year earlier he had been diagnosed as having chronic myelocytic leukemia. Tait, the father of five children, was a friend of McKinney and a neighbor of Gayneld Mackelprang, the school superintendent who also died of leukemia. The widows of these three men and the mother of Odessa Burch met with the law-yer, Dale Haralson, in June 1978.

Tait was a heavy-equipment operator. He worked for the local lumber company, helped build Glen Canyon Dam on the nearby Colorado River, and operated equipment for the town of Fredonia. He tired easily in the spring of 1964. A local doctor referred him to a specialist in Salt Lake City, who said he might live for another five or ten years with the leukemia. As his condition worsened, there were frequent trips back and forth from Fredonia to Salt Lake. Tait died in the hospital in nearby Kanab. His children had not known that his illness was terminal. Mrs. Tail went to work, and for the first time the family qualified for antipoverty programs.[64]

John E. Crabtree of Cedar City, the last leukemia victim, was a traveling salesman for a Salt Lake City hardware company. He had a wife and seven children to support. Crabtree took no vacations, and no sick time until his final illness, and he rarely left home when he was not out covering his vast territory. He enjoyed working in the yard. Crabtree was in his midfifties when he was diagnosed as having acute myeloblastic leukemia. Up until a few months before his death, he did not tell his wife that he had the disease.

During his last days in a Salt Lake hospital in early 1966, Crabtree experienced excruciating headaches. Painkillers did not help. His wife, Florence, suggested coffee and tea, since those caffeine-laden beverages might alleviate the pain. Crabtree objected at first. "I'd be breaking the word of wisdom," he said. He changed his mind. Mormon church officials at first hesitated, then relented and gave Crabtree permission to taste the forbidden drinks.

On the last night, Mrs. Crabtree left the hospital at 9:30. "That night," she said, "I told the Lord if he couldn't get better, to take him. I was just about to get off my knees from praying when my brother came in and he said, 'Florence, they've called from the hospital and said for you to get back there as fast as you can.' By the time I got there he was gone."[65]

Willard L. Bowler lived on a ranch about twenty miles north of St. George. He was active in the water district, the cattle association, and the Farm Bureau. The Bowlers' small ranch, about four miles from Veyo, Utah, consisted of a total of 1,000 acres on which they raised cattle feed. The cattle grazed about fifty miles southwest of the home ranch, which would put Bowler just about in a direct line from the test site to St. George.

Most of the day Bowler was outdoors tending his cattle. He undoubtedly was exposed to fallout, and a great deal of sunlight. Mrs. Bowler could remember seeing the beautiful mushroom-shaped red and orange clouds that faded off into duller hues after a short time.

In the spring of 1967, while rounding up his cattle, Bowler noticed that one of his fingernails had turned black. A doctor removed the nail, but the finger still bothered Bowler, so in August the doctor recommended that he have a biopsy performed. The diagnosis was a malignant melanoma, a type of skin cancer. The finger was removed in Salt Lake City. At first Bowler felt better, then he felt ill, as if he had the flu. Surgery and radiation treatments followed. Bowler returned home. He went into a coma. All five children, his grandchildren, a beloved brother, and Bowler's wife were around him when he died in December at the age of fifty-seven.[66]

Melvin O. Orton, a distant relative of Peggy Orton, grew up in Parowan during most of the years of atmospheric testing. He left in 1957 to go on a two-year mission to Nova Scotia for the Mormon church. On his return, he married and had three children. The family moved to Salt Lake City in 1962, and Orton held various jobs there, ending up as an auditor for the state.

In February 1970, at the age of thirty-two, Orton was operated on for a stomach complaint. Prior to the operation the doctor did not think that cancer was a possibility. As the doctor walked out of the operating room, however, he approached Juanita Orton, looked her straight in the eyes, and said of her husband, "Totally inoperable." The doctor then turned and walked away. More surgery and chemotherapy treatments followed. Mrs. Orton did not care if her husband became addicted to the drugs. His pain was intense. She urged the doctors, "Give him the medicine when he asks for it, regardless how close the doses are together." Orton died on February 11, 1971.[67]

In 1957 Marel D. Bradshaw, a miner, was working at the Nevada site when a test was conducted in one of the tunnels. Smoke and dust blew out of the top of the mountain. The only warning he had been given about safety was to abandon the construction camp when the siren blew. Thus he did not connect radiation with the later cancer deaths of three of his coworkers who had gone back into the tunnel on a recovery team. Bradshaw, his wife Delsa, and their four children lived in Pioche, Nevada, from where they could see the tests to the southwest.

In September 1972 *Delsa Bradshaw* was diagnosed as having a metastatic carcinoma of the lung, a form of lung cancer. She was a nonsmoker. Mrs. Bradshaw died in December at the age of fifty.[68]

Geraldine Thompson and her husband, Harold, played a game. Mrs. Thompson pretended that she did not know that she was terminally ill with ovarian cancer. Harold pretended that he was the only one who knew, and in this way the couple found that it was easier to keep the terrible secret from their three children. The children were told of their mother's illness about one week before Mrs. Thompson died in her Cedar City home in December 1973.

It was from their home that the family had watched the tests light up the early morning sky to the southwest. One day a pink cloud passed overhead. "We stood outside and watched it because nobody said there was any hazard," said Harold Thompson.[69]

Paul Wood also recalled a cloud with a pink hue, actually more a pinkish gray, he said, and more like dust than anything else. It wasn't anything like the normal clouds. Wood and his wife, Catherine, went outside their Cedar City home to watch the tests. The couple had five children. On November 30, 1972, *Catherine Wood* was diagnosed as having anular carcinoma of the colon. After the usual visits to doctors in Salt Lake City, Mrs. Wood died on January 23, 1974.[70]

As a baby, *Lisa Pectol* lived with her mother on her grandparents' farm in Veyo. One day her mother put her in a playpen and placed it in the doorway of the house. The mother could not remember whether the playpen was in the sun or the shade. Lisa, who was fair, sunburned easily. The months-old baby was clad in a diaper and a little white nylon frock. A fallout cloud passed overhead, a long, pink, dirty-looking cloud, and from it little sparkling particles descended to earth. When Lisa's mother went to check on her baby, she found that her daughter was burned. Her skin was white under the diaper, but the dress had not filtered out whatever was the cause of the burn. The government man, as Lisa's mother described him, came to the farm and said that there was nothing to worry about. The radiation from the cloud did not amount to anything more than what was on his watch dial, he said.

Twenty years later, Lisa was 5 feet 2 inches tall. She had blonde hair and blue eyes, and her husband, Dwight, thought she was very beautiful. He was building their home in Washington when Lisa, who was four months pregnant, began to suffer severe headaches and attacks of vomiting. At first the couple thought it was the pregnancy. The illness persisted. Lisa lost weight. She was admitted to a hospital on November 4, 1976, and died of a brain tumor three days later.

Lisa's sudden death and the death of their unborn child was a blow to Dwight Pectol. "It was as if my life ended," he said. "Never did another thing on my house. I just completely stopped. It was in the framing stage. I just couldn't do any more on it. I began to drink a lot. I didn't go to work. I didn't care if anything at all happened." This malaise lasted several months, or perhaps a year. Pectol was not sure. He lost forty or fifty pounds from his hefty frame and was almost unrecognizable. Gradually he phased back into the world, remarried, and was separated at the time of the trial.[71]

Louise Whipple was driving the school bus from Hiko to Alamo when she was stopped by the deputy sheriff on the morning of Shot Harry. Her youngest son, Kent, was on the bus with her. Twenty-four years later *Kent Whipple* died of adenocarcinoma of the lung, a form of lung cancer. Whipple fought hard to live; he had everything to live for.

Kent Whipple had spent most of his early life outdoors. He loved horses and the ranch life. Kent drank at least three glasses of raw milk a day and ate ranch beef and the vegetables grown in the family garden. He slept year-round on the screen porch, with his dog at the foot of the bed. In winter the snow drifted onto his bed, as did the fallout, on occasion.

The boy's father had died after a ranch accident some years earlier. Mrs. Whipple raised the children, ran the ranch, drove the school bus, and taught at the Alamo school, which Kent attended. During the early years of testing, the school received no warnings that fall-out was headed in its direction. Later, the AEC advised the principal when to keep the children indoors. Alamo was the closest downwind community to the test site.

Whipple left Hiko to go to a California college in 1957. He re-turned in 1961 with a wife after dropping out of college and serving an apprenticeship as a cattle buyer. Jane Whipple taught school for the first few years of their marriage, and Whipple worked hard to build up his cattle-buying business and his ranch. In time, Whipple began to prosper and Jane quit teaching to raise four children. The couple had an exceedingly happy marriage. A color photograph of the family looks like a glowing advertisement for western wear.

In the fall of 1975, Whipple felt tired and noticed that he was short of breath. He was normally a strong, healthy man. A local doc-tor sent him to a clinic in Santa Barbara, California, for a biopsy. The diagnosis was lung cancer. Whipple smoked an occasional cigar and drank infrequently. The diagnosis was within an acceptable period of time for the atomic tests also to be considered as a cause of Whipple's cancer.

When Whipple returned to Santa Barbara for radiation and chemotherapy treatments, he would not let his wife accompany him.

On the trips back to Hiko, the rancher remained in St. George until the nausea and other aftereffects of the treatments abated. He said nothing to his wife about these temporary illnesses. The couple did not tell the children that their father had cancer.

Whipple wanted desperately to live. He sought the advice of a palmist and a faith healer and made thirteen trips to a laetrile clinic in Tijuana, Mexico. The cancer kept spreading. Jane said that "the doctors told me, 'let him die, there's no use.' I said, 'You mean you just want to treat him like an old cow and drag him up the canyon, or something.'" Whipple was given more radiation treatments.

He never complained of the pain and took no medication for it. Near the end Whipple called out from his bed at home, "Jane, help me, help me. It's killing me. It's killing me. Help me." Jane said, "I didn't have a pill. It was midnight. I couldn't get hold of the doctors. I couldn't do anything for him, and it's the first time he asked for help. So I just crawled in bed and put my arms around him and held him, and he slept." (The courtroom was absolutely still, except for scattered movements to extract handkerchiefs. The judge sat sideways, with his head cast down.)

A few days later, Whipple died at the age of thirty-eight. He was buried in the Hiko cemetery, just across the road from the Murray Whipple ranch. "I mean, I just went numb," testified Jane Whipple Bradshaw. "No feelings whatsoever." She remarried, but that marriage was not successful. At the time of the trial she was about to lose the ranch that she and Whipple had purchased near Alamo. It was evident that Jane was still very much in love with Kent.[72]

Irma Wilson received social security payments after her husband died, and she also worked as a maid for the Dixie Medical Center in St. George. On this meager income she raised twelve children. In 1977 Mrs. Wilson was diagnosed as having cancer of the bladder, and she died on January 13, 1978. Like others in the downwind region, Mrs. Wilson had awakened her children in the 1950s to watch the early-morning fireworks in the sky west of their St. George home. Mrs. Wilson was a quiet person who was totally devoted to her large family. "She was a patriotic person," said her son, Delwin.[73]

James N. Prince did not notice the tests when he lived in Cedar

City. He was quite busy getting started as a young married man. In November 1977, his wife, *Daisy Lou Prince*, was diagnosed as having histiocytic lymphoma. She died nine months later at the age of forty-six. The couple had two daughters.[74]

Kenneth and *Donna Jean Berry* lived in Cedar City. He was a diesel mechanic and she worked as a telephone operator. They had five children. Kenneth Berry came down with diabetes and lost his eyesight. He could no longer work. The family then depended on Mrs. Berry.

On November 24, 1978, Mrs. Berry died of cancer of the ovary and uterus. The family collapsed after her death. Berry no longer wanted to live. He cried and cried, and sixteen days later he died. His heart just gave out, said the doctor. Two of the sons who were still living at home quit their jobs and withdrew into themselves.[75]

Glen Hunt lived in Cedar City and the small village of Hatch, fifty miles east, during the years of atmospheric testing. He worked on road construction and prospected for uranium. Hunt was diagnosed in August 1978 as having adenocarcinoma of the pancreas. Surgery, chemotherapy, and radiation treatments followed.

The pain was intense and Hunt was given pills every four hours. From 185 pounds, Hunt's weight dropped to 115 at the time of his death on September 19, 1980. He was survived by his wife, Sharon, and a daughter, Denise.[76]

The litany continued with the living, who could no longer look forward to a future with any certainty.

Jacqueline Sanders remembered seeing the tests, sticking her feet under a fluoroscope in a shoe store, and having dental X rays— all sources of radiation. She was five years old when the atmospheric tests began and could not remember having to remain indoors at school because of fallout. Of life in St. George during the fifties, Mrs. Sanders recalled that it was a small, friendly town, a place where one did not expect to be harmed and people trusted one another.

Mrs. Sanders graduated from Dixie College and went to work as a secretary in a local bank. At the age of twenty-one she was diagnosed as having adenocarcinoma of the thyroid. She was advised by her doctor not to marry. Mrs. Sanders did marry. She was advised not to have children. The couple, who moved to Salt Lake City, had six daughters. Mrs. Sanders was very conscious of the surgery scar on her neck. She had her hair cut in a certain way and wore high collars to hide it.

The family's activities were geared to her health. When her husband changed jobs, medical insurance that would cover Mrs. Sanders was a prime consideration. Her goal was to live long enough to raise her six children, the youngest of whom was fourteen months old at the time of the trial. The Sanders hoped to be able to return to St. George to live someday. Salt Lake City just did not have that same sense of closeness.[77]

William M. Swapp saw the flashes of light to the east and noticed a hazy sky over St. George after some shots. He was a heavy milk drinker. In 1971, at the age of fifty-two, Swapp was diagnosed as having a sarcoma of the kidneys. Now a kindergarten teacher in Logan, Utah, Swapp said that after surgery and extensive radiation treatments the doctors believed that there would be no recurrence of the disease.[78]

Jeffrey Bradshaw was born in Cedar City in November 1953. He could have been exposed to fallout from the Upshot-Knothole series while *in utero*. Then there were the later tests when he was growing up. In 1959 the family moved to St. George. Young Bradshaw was something of a phenomenon in that town. As a baseball pitcher, he never lost a Little League game until the state championships. He never lost a game in Pony League, and in high school he was on the all-state baseball teams and also played football and basketball.

Bradshaw's goal was to become a professional baseball player, and who knows, he might have made it. Bradshaw earned a baseball scholarship to Northern Arizona University, which in the year he would have been playing beat the national collegiate champions.

Bradshaw pitched batting practice in the fall of his freshman year at college. He felt quite tired. He got through that semester by resting in the infirmary during the weekends and studying hard so that he could retain his scholarship. A Salt Lake City cancer specialist diagnosed his illness as Hodgkin's disease in January 1973. Hodgkin's is a cancer that affects the lymphatic system. Although Bradshaw did not know it at the time, his dream of a big-league baseball career was over at the age of twenty.

Extensive radiation and chemotherapy treatments followed. The young man became addicted to painkillers and had to undergo a separate treatment for that problem. There was a brush with the law when he was charged with driving while drunk. The charge was dismissed when it was discovered that he was under the influence of medication. Bradshaw, who once had had great hopes, wanted to die.

The young man married in 1979 and went to work for his father, who owned the Ford dealership in town. Previous attempts to work elsewhere had not succeeded because of his illness. The couple wanted to adopt a child because Bradshaw was sterile from all the cancer treatments he had received over the years, but they could not because Bradshaw was not considered a good health risk.

Bradshaw, once the star athlete, hobbled about using a cane or crutches as his limbs deteriorated. His only exercise was swimming. He thought he was in fair condition at the trial, but others in his family thought differently.[79]

Lionel Walker remembered the clouds in the sky over Pioche. At the time, he was working in a mining mill and living on a ranch. He said, "They looked like water vapor from a cooling tower. You could see the clouds coming up behind the mountain and getting bigger and bigger. Airplanes were in and around them all the time they were in view. Generally by mid-day the thing was well past us."

Walker loved milk. He drank about three quarts of raw milk a day. He also used a Geiger counter to search for uranium. The hope then was that everyone was going to get rich from the radioactive material. There was also a great deal of cancer among Walker's blood relatives.

In 1955 Walker and his wife moved to the Seattle, Washington, area where he went to work for an oil company. The couple had four children. In March 1977, at the age of fifty, Walker was diagnosed as having adenocarcinoma of the prostate. The usual treatments followed, and Walker underwent an unusual amount of pain and discomfort. A good marriage was made difficult by his irritability and temporary impotence—the results of the treatments, which he considered worse than the disease. His career stagnated. Throughout the difficulties, Walker remained a staunch supporter of nuclear power plants and signed petitions favoring their construction.[80]

Norma Jean Pollitt lived in Cedar City during the time of atmospheric testing and could recall the dark, dusty fallout clouds and the constant assurances that there was no danger. Mrs. Pollitt—the mother of two children, a school teacher, and married to a teacher—was diagnosed as having adenocarcinoma of the breast in 1978. She was thirty-six years old.

What effect did that have on her plans for the future? She said, "Well, instead of living with the idea of a normal lifetime, I now live my life with the outlook that, 'Well, now today I made it.'" Mrs. Pollitt died a few months after the trial ended.[81]

CHAPTER 8

THE SCIENTISTS

FIRST THE TESTS, then the
cancers. Only the scientists possessed the knowledge to define the
cause-and-effect relationship, if any, between the two events. But
the loyalty of most of the scientists within the government, where
most of this highly specialized knowledge resided, was to institutions
and careers. Meanwhile, people died. The discipline of science,
which supposedly involves a systematic search for truth, was poorly
served during this time. The keepers of the flame were a hooded
priesthood.

The AEC scientists, with one noticeable exception, were be-
holden to policy. National security was the policy of paramount importance at the time, and it was a handy policy to have around.
Its invocation could serve the needs of defending the country while
also protecting jobs and reputations. These were the years of the
McCarthy and Eisenhower eras, when a maverick scientist was apt to
be branded a Communist and conformity was rampant. Under Chairman Lewis Strauss, who assiduously hunted security risks within the
AEC, a climate of fear pervaded the scientific bureaucracy. The Public Health Service officials within the Department of Health, Education, and Welfare were dutifully subservient to the more powerful
and better-funded AEC. They confined their protests to memos for
the files. Outside the government, the few dissenters, like Lyle Borst
and Ralph Lapp, were more vociferous but less credible, and the
academic community seemed more interested in obtaining federal
grants and publishing studies than in redressing wrongs.

That radioactive substances were dangerous to people's health was hardly a secret within scientific ranks. A few months after W. C. Roentgen announced the discovery of the X ray in December 1895, researchers who used it reported burned skin, hair loss, and impaired vision. Thomas Edison had sore eyes after being exposed to the radiation of an X ray machine in 1896. Around the turn of the century, researchers who were working with radium and uranium reported similar symptoms of radiation sickness. In 1901 a guinea pig was killed by X rays without leaving any visible marks on the animal's body. Radiation and leukemia in humans were linked in 1911, and between that time and 1959 more than two hundred cases of radiation-induced leukemia were mentioned in the world's scientific literature. By 1922 more than a hundred of the pioneers of radiology had died of cancer as a result of working with radioactive materials.

The year that the two atomic bombs were dropped on Japan, cancer research grants ranged from $500 to $1,500. Dr. Shields Warren, who would become the director of the AEC's Division of Biology and Medicine in a few years and go on to a distinguished teaching career at Harvard University, wrote in November 1945: "When we think that we can afford to spend two billion dollars and concentrate over a hundred thousand men to produce two bombs capable of killing or wounding over 300,000 human beings, the utter inadequacy of the present support of cancer research is only too apparent." Warren pointed out that the danger of radiation-induced leukemia was well known to radiologists. In fact, leukemia was nine times more frequent among radiologists than among other physicians.[1]

As the tests got underway at the Nevada site in 1951, it was known within the scientific community that radiation caused superficial injuries, leukemia, malignant tumors, genetic defects, and such miscellaneous ailments as cataracts, obesity, impaired fertility, and shortened lifespans. Also, an association had been found between the fallout from the Hiroshima and Nagasaki bombs and leukemia. A report issued by the Oak Ridge National Laboratory noted: "Preliminary indications reported from studies undertaken on the survivors of the atomic bombs at Hiroshima and Nagasaki in Japan suggest that there is a substantial increase in the leukemia incidence rate in the

exposed population over that found in a comparable control group."
But this information, as would be the case throughout the years of
atmospheric testing, was not passed on by the AEC to the downwind
population.[2]

What was not known with any certainty was the amount of radiation it took to cause these effects. Except for pregnant women, a
single dose of 25 roentgens was thought to be the threshold below
which no damage would occur. Yet, at best this figure was a guess,
since at the time there was no person with leukemia whose exposure
was known with any certainty. Controlled experiments had been
carried out on mice and other animals.

The threshold theory, on which the magic number of 25 roentgens is dependent, held that there was a safe level of exposure.
But through the 1950s and early 1960s the scientific community was
moving toward the linear theory, which held that there is no safe
level of exposure because no dose of radiation is so low as to eliminate any chance of inducing cancer. The less the exposure, the less
the risk. But there is no zero risk unless there is zero exposure. In
other words, there is relative danger to everyone. The risk of catch-
ing cancer from radiation is one of chance, since radiation that passes
through the body can do one of four things. First, it can do nothing.
Second, it can kill a cell, which is the goal of radiation treatments.
Third, radiation can damage a cell, but that cell may subsequently
repair itself. Finally, the radiation can damage the nucleus of the
cell, which may then multiply in its aberrant state and eventually
become a malignancy.

What this meant for the downwind population was that not everyone who was exposed to radioactive fallout contracted cancer, nor
could all those cancers that occurred in people who lived in the region during the years of atmospheric testing be attributed to fallout.
Just as assuredly, however, some could. Although more was supposedly known about radiation than about any other toxic substance,
huge gaps remained in the state of the knowledge.

The AEC was well aware of the danger of radiation to human
health by the late 1940s. Elaborate safety precautions were instituted in the national laboratories that were not available or made

known to the downwind population. K. Z. Morgan, director of the Health Physics Division at Oak Ridge National Laboratory, wrote in 1949:

> Every precaution is taken in the laboratories to keep these exposures at a minimum. The health physics surveyors are continually on the job, limiting the working time in the various areas so that external exposures are kept very low. All persons working in areas where radioactive exposure is possible wear pocket meters and film meters that measure the day-to-day and accumulated weekly exposures, respectively. If by accident a person receives an appreciable fraction of a permissible daily or weekly exposure, he and his supervisor are notified immediately, so that he will rectify conditions causing the exposure and so that he will not repeat these exposures. Health physics surveyors keep a constant check on continuous air monitors and water monitors, in order to maintain the levels of airborne and water-borne activity far below those considered to be the maximum permissible concentrations.[3]

At the Nevada Test Site, blood tests and urinalyses were performed regularly, and similar precautions were taken when tests got underway. But nothing was done for the offsite population, except to issue hit-or-miss warnings to stay indoors, wash cars on occasion, and in later years to hand out a few film badges. No mention of cancer showed up in the AEC's many public statements, press releases, films, or booklets. There seems to have been a blackout on that dirty six-letter word.

Yet the possibility of injury and death to the offsite population was being discussed within the AEC. Project Gabriel, first undertaken by the AEC in 1949 and then updated in 1952, attempted to assess the long-range effects of fallout. The best scientific minds within the AEC were put to work on the project. In terms of a one-time crop contamination, the 1949 report from Shields Warren to the commission stated: "It is obvious that a determined people *aware* of the danger could either migrate or obtain food from other sources" (emphasis added). In 1952, Warren made an interesting observation at the end of his updated report. Referring to the effects on human health of a rainout, such as would occur over New York State and St. George after Shots Simon and Harry the next year, he told the commission: "Under extreme weather conditions, the wash-out from a 13 KT bomb would be potentially lethal to a large fraction of the inhabitants in an area of considerably less than 100 square miles (taking 400 r as the mid-lethal dose)." Shots Simon and Harry were considerably larger than 13 kilotons. The first report was classified top secret; the second, secret.[4]

The effects of a rainout were a fearsome prospect indeed. The Project Gabriel estimates for a rainout and Harold Knapp's later estimates for St. George were not that far apart. Knapp, a former AEC scientist, estimated that if there were a rainout over St. George from Shot Simon, which dumped fallout on New York State under just such conditions, residents of the town would receive a dose of perhaps 453 rads, with a low of 101 rads under certain conditions and a high of 909 under others. Scattered rain did occur over St. George after Shot Harry. Had the fallout from Shot Harry coincided with a full-blown thunderstorm, such as that which occurred over New York: "It could have literally killed half or all of the people in town," Knapp testified at the Allen trial.[5]

The fact that cancers might occur in future years was brought to the attention of the test site management and other AEC officials by Howard L. Andrews, head of the radiobiology section of the National Cancer Institute, after the Upshot-Knothole series. The AEC brushed it aside. Andrews also wrote an article on fallout in a September 1955 issue of *Science* magazine in which he devoted far more space to genetic effects than cancers.[6] In the 1950s the scientific community thought that risks to the genetic structure were greater than risks to the body. More recently, the pendulum of scientific fashion has swung the other way.

Rather than a systematic search for harmful radioactive isotopes, there was a reactive quality to the concerns voiced about them that verged on faddishness. The radioactive isotopes strontium 90, carbon 14, and cesium 137 were given a great deal of attention in the

1950s, while iodine 131 only came into its own in the 1960s as the result of a bitter feud between technocrats within the AEC. The two protagonists were Gordon M. Dunning and Harold A. Knapp, Jr.

It would be difficult to find two scientists with such opposite temperaments (Dunning was dour, Knapp was ebullient), which may be one reason why they clashed so heatedly. But the significance of their feud transcended character traits. Their differences were indicative of how unwanted evidence that fallout causes cancer was treated within the AEC.

Gordon Dunning, who has appeared as an omnipresent shadow throughout this account, dealt with the issue of local fallout from the time he joined the AEC in 1951 until his retirement in 1972. By the mid-1950s he was the commission's foremost authority on the subject. To the plaintiffs' lawyers in the Allen suit, Dunning was the chief villain. Dale Haralson said, "Regardless of the motivation, the consequences of his conduct were regrettably tragic and unjustifiable."[7]

To Haralson and the other attorneys, Dunning was a mystery man. They were told by an informant, a former high-level official in the Department of the Interior during Stewart Udall's tenure as secretary of that agency, that Dunning may have been in the employ of an intelligence agency. The informant accompanied Dunning on a survey of Amchitka Island in the Aleutians, which was being considered as a test site, and noticed the considerable influence the scientist had at the various military bases along the way. Asked by Udall at a pretrial deposition interview whether he had ever worked for or had any contacts with the Central Intelligence Agency, Dunning replied that he had not.[8]

The government lawyers at the Allen trial depicted Dunning as a loyal servant of the AEC and his country. Richard G. Hewlett, the AEC historian at the time, remembered Dunning as an extremely busy, methodical scientist. "He had a variety of roles which were not well defined," said Hewlett. "He was looked to for advice because of his abilities." Although Dunning was supposed to report to the head of the Division of Biology and Medicine, he also served the AEC commissioners directly. Dunning shuttled back and forth between AEC headquarters, the national laboratories, and the Nevada and Pacific test sites, where he accumulated a small dose of radiation.[9]

Dunning had an undistinguished scientific background. After graduating from State Teachers College at Cortland, New York, where he had majored in physical education and biological sciences, Dunning worked as a schoolteacher in the late 1930s. After World War II he obtained a master's and a doctoral degree in science education from Syracuse University. He taught at the college level and then went to work for the AEC as a biophysics research analyst. Other jobs that Dunning held within the Division of Biology and Medicine during the years of atmospheric testing were biophysicist, radiation effects specialist, and then chief of the branch that dealt with the radiation effects of atomic weapons. He was a frequent visitor at the Nevada site, where he first served as a representative of the Division of Biology and Medicine and then was on the test manager's advisory panel.[10]

It was Dunning's technical mastery of a highly complex, potentially hazardous, and politically sensitive subject and the access he had at the field, administrative, and policy-making levels of the AEC that made this middle-level bureaucrat so powerful. One way to survive and prosper within a bureaucracy is to master a complex and important but narrowly focused subject. Dunning did just that. Bosses came and went, but the technician remained. His role was that of advisor, cajoler, interpreter, note taker, and when it became necessary, decision maker. That Dunning's role could be defined so well is a tribute to the extensive paper trail he left behind, most of which was classified at the time.

In earlier years, he had been short, trim, and had had wavy blonde hair and a thin mustache. His colleagues thought he was quite debonair. At the trial, Dunning, who was in his seventies and living in retirement in Arizona, displayed that same sense of trimness and self control, but his hair and complexion were grey. He was suffering from cancer. That was about all that Gordon Dunning and the plaintiffs had in common.

There was one documented moment when Dunning lost his scientific cool. As one of the first scientists to recognize the possible harmful effects of iodine 131, he oversaw the collection and analysis of milk samples from around the test site. Morgan Seal of the Public Health Service developed a method of estimating the iodine content of milk. He showed the results to Dunning, who exploded. Seal said, "He got mad, red in the face, took it and threw it on the floor and stomped on it. 'Don't you do that.' So I don't know whether it meant a darn thing or not. It is immaterial, but it sure got Gordon excited." Seal and Dunning later said that the disagreement was over the technique of analysis, not the results.[11]

By writing for scientific journals, Dunning obtained an important audience for AEC views on radioiodine. He wrote in a 1955 issue of the *Scientific Monthly* that "the highest measured radiation exposure to the thyroid of human beings has been far below that needed to produce any detectable effects." The key word is *detectable*, since it would take a number of years for thyroid cancer to appear, due to its long latency period. The six-page article, "The Effects of Nuclear Weapons Testing," did not mention cancer. Another article by Dunning that same year in the *Journal of the American Medical Association* contained many of the same reassurances and the same absence of pertinent information. Cancer, again, was not mentioned. The article was reprinted and distributed by the AEC.[12]

The popular media also offered no real enlightenment at this time on the effects of fallout. During the next two years, when the fallout issue heated up and climaxed in congressional hearings, the public was bombarded with contradictory evidence and trivia. Again, the words *cancer* and *leukemia* were absent from or downgraded in almost all the accounts.

A *U.S. News & World Report* interview with Dr. Robert H. Holmes, director of the Atomic Bomb Casualty Commission, brought the headlined news that the survivors of the Hiroshima and Nagasaki bombs were, "for the most part," leading normal lives. It was explained, but not elaborated on in the body of the text, that the relationship between the commission and the AEC was "contractual." Buried farther down was the remarkable but obscure statement that there had been a "significant" increase in leukemia among the survivors. The magazine, and Dr. Holmes, hastened to add that the increase amounted to only 103 cases.

Life magazine contributed picture spreads of fireballs, a silhouetted sheriff, an irradiated test site guard, and irradiated burros at Oak Ridge, sheep at Hanford, and beagles at the University of Utah. Plastic protective masks on bronzed busts were juxtaposed against a picture of "worried senators" at the hearing. There was a picture of a balding Alvin Graves testifying. On another page were the Bordolis, saddened by the death of their son, and "an embattled rancher," who evidently represented a species.

Such diverse magazines as *Newsweek, Time, McCall's,* and *Better Homes and Gardens* contributed to the bewildering cacophony of voices directed at the public in an age when magazines and newspapers were the primary sources of entertainment and news. *Better Homes and Gardens,* in an attempt to be original, coined the phrase "bomb-dust radiation," meaning fallout. It was shortly before the 1957 hearing that Paul Jacobs made his thoughtful contribution in the *Reporter.* With this single exception, journalism had failed to perform responsibly on this issue. As with the other institutions involved, it was not the media's finest hour.[13]

Dunning's most visible public moment was when he testified for the AEC on local fallout at the 1957 hearings of the Joint Committee on Atomic Energy. The hearings, which were concerned mainly with global fallout, were tailor-made for the testimony of AEC bureaucrats like Dunning. Witnesses were restricted to "persons of competent scientific background who have been actively engaged in studying the various aspects of the fallout problem." The joint committee bent over backward to give the AEC its day in court. The director of the Division of Biology and Medicine was selected to be the leadoff witness because of his objectivity, said the committee chairman.[14]

When he appeared before the committee, Dunning did not participate in any discussion of cancer, at least none that a lay person would recognize as such. He briefly touched on the effect of strontium 90 on bone marrow but never mentioned the word *leukemia.* He testified that no offsite residents had suffered radiation burns.[15] That statement was, at the very least, an overreliance on the AEC officials recognizing radiation burns as such.

After the hearings, key AEC staff personnel met for an in-house assessment of the testimony. They agreed that the hearings had "generally progressed quite well, although the majority of the news reports featured the danger of fallout or stressed the scientific disagreement which exists." With disarmament talks underway, it was decided that the public relations program "will be basically directed toward demonstrating the relatively small hazards associated with testing and showing that we must move ahead with weapons testing until a test ban agreement is concluded which provides adequate safeguards." This was the AEC policy that Dunning, who was not present at the meeting, and others followed even after the test ban treaty was signed six years later.[16]

In May 1959, Dunning made a presentation to the joint committee on the estimates of the average cumulative doses of gamma radiation from fallout produced during the seven years of atmospheric testing at the Nevada site. The estimates for the downwind communities were put together by a committee of six scientists, all of whom worked for the AEC or were under contract to it. The committee was headed by A. Vay Shelton of the Lawrence Radiation Laboratory. Its technical work was done by Dunning, and the results became known as Dunning's estimates.

The average estimates for all the tests were, in some cases, lower than the calculations made immediately after a single shot, because of a complex formula Dunning and others in the AEC devised that took into account factors that diminished the dose, such as shielding, the weathering of isotopes, and the biological repair of cells.[17] These AEC estimates were on the low end of the scale. They were later used widely by PHS researchers, most of whom believed they were accurate.

The AEC estimates of the average dose of external gamma radiation for locations mentioned in this text were, from west to east: Nevada—Tonopah, 1.08 roentgens; Fallini ranch, 1.98; Bordoli ranch, 2.04; Lincoln Mine, 5.95; Groom Mine, 4.90; Whipple ranch, 1.10;

Alamo, 1.39; Caliente, 0.76; and Pioche, 0.74. Utah—Santa Clara, 4.27; Veyo, 2.82; St. George, 3.70; Washington, 3.30; Hurricane, 4.35; Cedar City, 0.64; and Parowan, 0.42. There was no estimate for Fredonia, Arizona. The estimate for the nearest community, Kanab, Utah, was 1.62.[18] Dunning and Knapp were about to collide on the issue of numbers, and Knapp would be at the high end of the scale.

While Dunning was the organization man par excellence, Harold Knapp was the maverick—the odd man out in the organization. After obtaining his doctorate in mathematics with a minor in physics from the Massachusetts Institute of Technology, Knapp went to work in the Office of the Chief of Naval Operations on special warfare problems. He was an operations analyst, a job that Knapp described as asking the right questions and then finding the answers. A necessary prerequisite for such a profession, said the analyst, was "innocent curiosity," of which Knapp possessed an abundance.

Knapp joined the AEC in 1955 and five years later became part of the newly formed Fallout Studies Branch within the Division of Biology and Medicine. During the early 1960s Knapp was involved in the successful attempt to free three black men who had been sentenced to death for the supposed rape of a white teenager near Knapp's home in the rural Maryland countryside. The AEC scientist's persistent curiosity led him to ask the right questions that eventually led to their release from jail. For that activity he received the Oliver Wendell Holmes Award from the American Civil Liberties Union in the capitol area.

In 1963, at the height of his struggle with Dunning, Knapp left the AEC and went to work for the Washington-based Institute for Defense Analyses, which did highly sensitive studies on nuclear warfare for the Joint Chiefs of Staff, the Office of the Secretary of Defense, and the Defense Nuclear Agency. Among his many classified studies while there, Knapp dealt with the survivability of the president and his successors during a nuclear war. In 1981 he joined the Joint Program Office in the Department of Defense, whose innocuous title hid the awesome responsibility of designing and putting into effect a system that would assure the continuity of government during nuclear war.[19]

When Knapp was served with a subpoena to testify in the 1982 sheep trial, his superior, a retired army general named Richard G. Stilwell, said, "Jesus Christ, Harold, testify against the government? We'd better get that quashed in a hurry on national security grounds." The subpoena was not quashed, and the general subsequently read an account of Knapp's testimony in a Virginia newspaper under the headline "Defense Official Charges 'Conspiracy' to Conceal 1953 Atomic Test Dangers." Knapp lost his job as deputy director of the office. A few months later, at the time of the Allen trial, Knapp was at work in another Department of Defense office that dealt with the problems of maintaining communications between the leadership and the military forces in a nuclear war.[20]

Knapp was a paradox. He was highly placed within the nuclear establishment yet very sympathetic to the victims and quite helpful to their attorneys. During the Allen trial, Knapp scribbled a note to the four lawyers: "I only regret that I have but one life to give to the plaintiffs." He insisted on being issued a subpoena so that he could show his superiors that he was forced to testify against the government. Knapp, in his late fifties, had a foot in both worlds. He was the owner of a Christmas tree farm in New Hampshire and had a fallout shelter in the basement of his home.[21]

Knapp and iodine 131 came together because of his curiosity. His first assignment in the Fallout Studies Branch was to review a report that had been prepared for Senator Clinton P. Anderson, chairman of the joint committee. He thought the report had been poorly done, so he redid it. The revised report briefly mentioned doses of radioiodine to the thyroid that children had received in St. Louis. One report led to another; there were subsequent studies on strontium 90 and external gamma radiation. Knapp then reconsidered the relationship between iodine and the relatively high doses he estimated for St. Louis. A question occurred to him: "For heaven's sake, what must the doses have been around the Nevada Test Site? That simple thought led to all of the effort and tremendous controversy over the radioiodine question, which had been totally overlooked for ten years," said Knapp, who then set out, like the eternal innocent, to find the answer.[22]

Dunning warned Knapp that he was "playing with dynamite" when he took on the question of radioiodine around the test site, but the operations analyst persisted. Knapp had difficulty in getting information from Dunning for his study, and at times he received "angry, discourteous responses" from Dunning, who five years earlier had written an article stating that iodine 131 levels around the test site posed no health problems. That Knapp—who was the new, rather obstreperous kid on the block—might contradict that finding did not endear him to Dunning. Also, what Knapp was on the track of would show what the AEC had overlooked. While concentrating on guidelines for external gamma radiation, it had overlooked the fact that internal doses could be crippling or lethal.[23]

Knapp submitted his radioiodine report, which suggested that there were high radiation levels around the test site, in September 1962. A covering letter voiced the hope that it would be widely disseminated for review both within and outside the government. Knapp wrote: "If the conclusions are incorrect or misleading at this stage, then they can be blamed on me; if they are essentially correct we at least have the advantage of telling unpleasant news ourselves, and we are not vulnerable to a charge of having suppressed or misrepresented information on fallout, or worse still, of not being competent to find our own problems." Knapp was, indeed, quite naive. The report never left the precincts of the AEC.[24]

A little more than a month after its submission, Charles Dunham, who had taken over as director of the Division of Biology and Medicine, wrote Knapp a scathing letter. Dunham said the report was "amateurish to a degree" and contained an "almost total absence of thoughtful basis for many sweeping assumptions." What was needed in the new version was "detailed critical analysis by the author of each and every assumption made." Knapp, after a trip to the Nevada site to gather more information, rewrote the report. He made some changes, but he came to essentially the same conclusions with better documentation.[25]

Gordon Dunning reviewed the second draft and made some technical objections. He then raised a question of policy in a backhand manner by stating that someone else had posed the question: "In response to a question asked as to the motive for publishing the paper, a member (not the author) of the Commission staff said, 'The Commission has been telling the world for years that it has been conducting its operations safely, now it appears that this may not be so.' If a member of the staff says this about the paper, what reaction may we expect from the press and the public?"

Knapp shot back, "I expect somebody (not the author) might want to hang Dunning from a sour apple tree." [26]

The AEC convened a special review committee to determine the fate of Knapp's paper. One of the members of the peer review committee was John Gofman, an associate director of the Lawrence Livermore Laboratory in California. Dunham telephoned and asked him to come to Washington.

Gofman asked why. "I can't tell you on the telephone," said Dunham.

None of the five members from various universities and AEC facilities around the country knew the purpose of the meeting until they met on April 2, 1963. The chairman was Wright Langham, Group Leader for Biophysical Research at Los Alamos.

Dunham explained the background and major findings of the report. Knapp had found that the dose to the thyroid was about one hundred times higher than previous AEC estimates. The committee members were still puzzled as to why they had been called to Washington. Dunham continued, "If Dr. Knapp publishes his findings, the public will know that they haven't been told the truth up to now, and my request to this committee is to see whether you can convince Dr. Knapp not to publish his findings." The committee was given a copy of the report to read, and Knapp was permitted to enter the room. Gofman said, "He was a bit angry when he came into the room, and I must say were I in his position, I would have been even angrier to have a committee called to look at your findings in that manner." Knapp answered questions and departed. Dunning entered the room and told the committee his opinion of the report. The committee advised Dunham that the report, with some modifications, should be published. "We thought the sky wouldn't fall if the truth were to come out about this," said Gofman. [27]

Langham tried to moderate the internal dispute while in Washington. When he returned to Los Alamos he counseled Knapp in a fatherly tone to rewrite the report in a straightforward manner without including any speculation. "It is my feeling that we cannot say unequivocally that the concepts can be applied to past situations," wrote Langham. "Even if they could be, nothing much could be done about it except add to the great public and political controversies raging around the concept of strong nuclear deterrents."[28]

The committee's approval perplexed Dunning. He wrote that

> we are in a dilemma. An ad-hoc committee reviewed the paper and, although I have not seen a statement of their conclusions, apparently they felt that the paper (with editorial rewriting) could be published. One might argue that the committee was not made fully aware of the pitfalls inherent in the treatment given the field data. But the fact remains, they did not turn thumbs down. Now what do we do? I wish I had a brilliant answer to the question. I do not.

There was one possible "out" for the commission, said Dunning. He suggested, "Let the Commission tell Dr. Knapp in a matter-of-fact and bland manner that the Commission interposes no objection if he, as an individual scientist, wishes to publish his paper." No, said Knapp, "It wouldn't work. Dr. Knapp might find out about this suggestion and go up in smoke. He is said to be mighty sensitive by now to this type of scheming."[29]

Tired of the unpleasantness, Knapp left the AEC. Dunham, his superior, wrote soothingly, "I shall miss the dogged persistence with which you pursued the problem of your choice." He added, "And as your parting shot your efforts to aid in explaining the vagaries of I-131 will pay off, I am certain, in a much better 'control' of the hazards associated with weapons testing."[30] Knapp replied, "I'm not unaware that there were times—probably most of the time—when you must have found me totally exasperating. If it helps at all, I get mad at myself, too."[31]

Knapp fretted over what the AEC would do with his report. Langham counseled him to sit back, relax, and get some "satisfaction out of seeing just how important you are when it comes around to stirring up a hornet's nest." Langham said that he "was having a little fun and am developing a considerable interest in the whole affair."[32]

The "affair" blossomed of its own accord. The commission was in a bind. The existence of the report was known outside the AEC. Failure to publish it would result in charges of suppression. That accusation was less desirable, in a public relations sense, than publishing the report in a routine manner and hoping it would be ignored. The AEC commission approved publication of the report along with the review committee's critique of it—a compromise and an unusual procedure.[33]

Again, Dunham was quite solicitous—compared to his first scathing criticism of the report—when he advised Knapp of the commission's approval. He did not want that particular volcano, who was now independent of the AEC, to erupt. Dunham wrote, "In my opinion, the combined package should give interested people something very worthwhile to chew on." To the commission he presented another case for publication. "Failure to publish a report of the study might be looked upon as a deliberate action by the Commission to withhold 'damaging information,' information for which the public has a genuine need. If such a reaction were engendered, an even less satisfactory public relations situation would ensue than would be created by publication in some routine manner." The commission opted for the lesser evil.[34]

The committee's comments were included in the front of the report, which was entitled *Iodine-131 in Fresh Milk and Human Thyroids Following a Single Deposition of Nuclear Test Fallout.* The comments dealt mostly with the sketchiness of the data on which Knapp had based his calculations. The report was published, said the committee, to "stimulate the scientific community's interest in the iodine-131 problem, which may be potentially quite important to civil defense and peaceful uses of atomic energy." It was also of great importance to the downwind population, but that was not mentioned. Nor was it very likely that such a technical publication would circulate as widely in that region as the film or pamphlets had.[35]

In order to get the report published as an AEC technical document, Knapp had to make one further concession. The estimated doses to the thyroids of children from Shot Harry, which ranged from 120 to 440 rads, were left out, though they were included in a condensation of the AEC article published in the British scientific journal *Nature* in 1964.[36]

K. Z. Morgan, often called the father of the health physics profession, knew the participants and observed the feud and its outcome from the remove of Oak Ridge. At the trial he said of Knapp, "I was very much put out at what I felt was the treatment he got in that period." To Morgan, Dunning was an uninspired scientist who was "the chief defender within the AEC of the 'no harm' theory."[37] Morgan wrote two long letters to Knapp in the summer of 1964 concerning the radioiodine controversy. The letters are remarkable documents because they reveal, without guile or the pressures of the witness stand and bureaucratic politics, how an eminent scientist viewed the morality of the situation at a time that was closer to the actual events. The letters go right to the core of the issue. Like Dunning and others within the AEC, Morgan thought it was a case of benefits versus risks, and some lives might have to be lost in the process. What he could not tolerate was the deception. Morgan wrote:

> I am not so much concerned that some of our population may have received high doses of radiation or that some children may die of thyroid cancer or leukemia as a consequence of dose from weapons fallout. I think in these cases one must weigh the benefits against the hazards and that very probably the benefits from testing weapons and assuring ourselves that we have nuclear devices that properly function is worth a number of lives. I am not competent to make these judgments but am quite willing as a health physicist and as a citizen to accept the fact that injuries and a few deaths from weapons fallout must be accepted in the best interest of our country.
>
> My concern is with the fact that there seems to be a tendency to repress information. I can't help but ask the question, "From whom are we hiding such information?" It seems to me

that in a democratic society that the people are the government, and they have a right to the facts and/or all available information of an unclassified nature that may enable them to arrive at their own opinions and conclusions. Certainly the fact that these weapons tests may have resulted in damage to some persons is not a classifiable piece of information. Of course, we would like to have more data, and data with lower uncertainties, but this is no reason for withholding information that might stimulate further investigations and procurement of the facts. Certainly the purpose of our government is not to protect its agencies and to guarantee that these agencies never make a mistake, but rather that these agencies serve the people in their best interests. If some people in the United States have been damaged—and certainly some have been if large numbers of children received doses from 20 to 100 rads to the thyroid—it is a pressing obligation of our government to compensate these people and to provide them with the best possible medical care.[38]

On the integrity of the AEC scientific staff, Morgan wrote, in the second letter:

I am sure you realize it would be difficult for me to comment on this subject in a letter sent under a United States stamp. It goes without saying that we expect there are some dishonest people in all large groups and in any government agency. I regret to say that on many occasions I have been brought to the conclusion that the principal objective of many government employees is to prove that they have not made a mistake rather than readily admit a mistake and then proceed to make amends.[39]

Eighteen years later at the Allen trial, Morgan, over the heated objections of a government attorney, was asked what he would have recommended at the time. He replied:

I would have given the same advice as I do now to community groups. That we have radiation with us, we've always had some

radiation with us, and it's an added risk. The people deserve
to know what the risk is, and they should make the decision on
how much radiation they want to take or are willing to take,
and there shouldn't be any cover-ups. There shouldn't be any
withholding of information. That the people themselves should
decide on their own future and the future of their families.[40]

Asked how the various government committees that set radiation
standards were established, Morgan, who was the recipient of the
first gold medal from the Swedish Academy of Sciences for his work
in radiation protection, said of those he had served on, including the
principal national and international committees:

I would like to be able to tell you that they're established in
a very scientific and always in an extremely proper manner.
I'm not sure from my experience that I can take that position.
Very often those that want the information or want the decision also are quite set on what answer they would like to have.
So human beings being as they are, and government agencies
and private industries being what they are, when they select
groups to make a decision it turns out that very often those
parties already have made certain decisions in their minds.
Those choosing these committees very often knowingly or pur-
posefully or by accident choose those that they believe will
favor the decision that they seek.[41]

There is a case in point. In 1962 the Federal Radiation Council
Working Group, of which Dunning was a member, was beginning
to put together a report on radiation standards and human health.
The executive director of the council, Paul C. Tompkins, who was
also deputy director of the AEC's Division of Radiation Protection
Standards, reported to an AEC commissioner that the group would
start by drawing up the conclusions. "We would then identify the
major questions that could be expected to be asked in connection
with these conclusions. It would then be a straightforward matter to
select the key scientific consultants whose opinions should be sought
in order to substantiate the validity of the conclusions or recommend

appropriate modifications."[42] Asked what he thought of such a procedure at the trial, Morgan said, "I was appalled that he would put anything like this in writing."[43]

Meanwhile, as a result of Knapp's efforts, public attention focused on the dangers of radioiodine. The Joint Committee on Atomic Energy held four days of hearings in late August 1963. Dunning testified and downplayed the dangers. Asked if he knew of any damage to the thyroid caused by iodine 131, he indicated there had been none. Although the question was framed in terms of "any damage," conceivably it was answered only in terms of the fallout from the Nevada tests. Dunning was well aware that radioiodine, ingested via milk and other foods, caused thyroid cancer, particularly in children. He had written about it, had attempted to quash Knapp's efforts, was aware of a second study done at the University of Utah that differed with the AEC's position, and was keeping an eye on the PHS monitoring network around the test site, which was searching for just such isotopes. In fact, the PHS cleared its press releases on what they found with Dunning.[44]

The AEC line was echoed at the White House. Asked about the Knapp report at a news conference during the hearings, President John F. Kennedy labeled it "controversial" and went on to give an equivocal answer that was full of "buts." He said:

> I don't think we should mislead the people there, that there is evidence on hand of a serious deterioration there. But, of course, it is a matter of concern to us that we not continue. But we are looking into it. But I would say that as of now that we do not believe that the health of the children involved has been adversely affected. But it does tell us, of course, these matters require further study.[45]

Kennedy was referring to a study undertaken shortly before the hearings by Gofman at the Livermore laboratory. The AEC hoped this in-house study would diffuse the radioiodine controversy. Instead, it added fuel to the fire.

Gofman said that before taking the job he had been assured by AEC chairman Glenn T. Seaborg: "All we want is the truth, Jack."[46]

One of the first persons Gofman hired was one of his former graduate students, Arthur Tamplin, who was made leader of the Information Integration Group.[47] Tamplin began to work on radioiodine.
His group reviewed all the available written material, including the
reports of the offsite monitors, and they began to come up with some
numbers of their own. Gofman briefed the director of the laboratory,
Michael M. May, on Tamplin's progress in late November 1965. The
situation did not look good for the AEC. A disturbed May wrote
Seaborg that "the possibility remains that the findings in St. George,
Utah, and surrounding area will get worse instead of better."[48]

At Seaborg's request, Gofman and Tamplin submitted an interim
report to Dunham, the director of the Division of Biology and Medicine. The estimate of the radiation dose to the thyroids of children
in St. George who drank one liter of fresh milk per day was 1,200
rads, while those in Salt Lake received 100 rads. The number was
scaled down after the Livermore scientists talked with Dunning in
Washington. Dunning read the interim report and experienced deja
vu. He said it would "do more harm than good to initiate the re-
opening of Pandora's box." Any finished report should be seen by
the AEC. "So we go around the merry-go-round again," Dunning
commented.[49]

The final report, entitled *Estimation of Dosage to Thyroids of
Children in the U.S. From Nuclear Tests Conducted in Nevada During 1952 Through 1957*, was published in May 1966 as a Lawrence
Radiation Laboratory document. Prepared under the auspices of the
AEC and reviewed by the commission, it was the best indication
yet of the nationwide spread of radioactive fallout from the Nevada
Test Site. The pattern of thyroid doses, as calculated by Tamplin and
his group, showed the highest concentrations in a wide band from
the test site southeast to the New Mexico–Texas border, moderate
doses more directly east from the middle of Colorado to the western
borders of Missouri and Arkansas, low doses through the balance of
the Midwest and south of New England, and moderate doses in a
band from upstate New York through Massachusetts. A sampling of
the estimates of radiation doses to the thyroids of children included:
St. George, 120 rads; Roswell, New Mexico, 57; Salt Lake City,

46; Grand Junction, Colorado, 33; Amarillo, Texas, and Boston, 19; Albany, 15; Des Moines, Iowa, 9; Cleveland, 8; and New York City, 5 rads. For the two years of atmospheric tests not included in the report (1951 and 1958), an increased dose of no more than 20 percent could be expected for most locations. A few were over that amount.[50] The Gofman-Tamplin effort represented a middle road in the numbers game, and it clearly tied the remainder of the country to the immediate downwind region.

It was difficult at times to determine what the word *Public* was doing in the title of the United States Public Health Service. Rather than serving the public, the agency was the handmaiden of the AEC.

The AEC had maintained countless times that there was no danger from the tests, but an awareness that leukemia and thyroid cancers may have been caused by the atmospheric tests converged during the early 1960s upon the AEC via the activities of the PHS. These were the chickens coming home to roost that Howard Andrews of the National Institutes of Health had warned about a decade earlier. The solution was to close the door of the chicken coop— in other words, to hush up the PHS. In return, the PHS relied on Gordon Dunning's low dose estimates, and in this manner no untoward conclusions were reached.[51]

In 1961, Edward S. Weiss of the Division of Radiological Health within the PHS undertook an epidemiological study of leukemia deaths in Washington and Iron counties, in which St. George and Cedar City were located. The study consisted mainly of reviewing death certificates and hospital and doctors' records to ascertain the number of leukemia deaths in the two counties between 1950 and 1964 and then relating that figure to the national rates and Brooklyn, New York, which happened to coincide. It was an invalid comparison that lowered the results. Utah had a much lower cancer rate than the rest of the nation because of Mormon abstinence and a predominantly rural setting—a fact that Weiss was aware of. The study was flawed.

Twenty-eight leukemia deaths were found for the period, versus an expected number of nineteen that was drawn from the Brooklyn study. There was an "anomalously high frequency" of six deaths in 1959, six years after Shot Harry and within the expected peak period for radiation-induced leukemia. As with the sheep deaths, no cause other than residency in the area was given for the excess number of human deaths. The 1965 Weiss report concluded: "Beyond the fact of extended residence in the area, there is no evidence to associate these cases with fallout exposure, other environmental contaminants or familial disabilities."

The five-page study was, however, the first indication that there was an excess number of leukemia deaths, albeit a small number, downwind from the test site. In arriving at his conclusion on fallout, Weiss used Dunning's dose estimates. Clark W. Heath, Jr., chief of the Epidemiology Branch, was asked to review the Weiss report, and although he found certain shortcomings in it, he recommended that it be published. "The question was of a lot of interest," said Heath. "I felt it was a fairly reasonable work-up—not overstated in terms of conclusions."[52] Although Weiss and others within his agency hoped that the study would be published, it never was. The AEC had other ideas.

The four leukemia deaths in Fredonia next drew the attention of PHS investigators, who were primarily interested in finding a viral cause for leukemia and were probing such clusters on a national basis. On the face of it, the four deaths in the small Arizona town looked unusual, so the PHS launched an investigation. Blood samples were taken from members of the four families, and they and other members of the community were interviewed. The study began in late June 1966, and the cursory three-page report was submitted in early August. No viral cause could be found for the unusual number of leukemia cases—approximately twenty times the number that might normally be expected for the small town. Actually, the figure was even higher, because the PHS again compared Fredonia, which was predominantly Mormon, to the national average.[53]

It was as if the very mention of radiation as a possible cause of cancer was an institutional taboo. The PHS made no independent assessment of radiation doses, accepted Dunning's estimates without assessing them, and barely mentioned fallout when questioning the Fredonia families. The government doctors were told by the AEC that the radiation levels were too low to cause leukemia. They meekly accepted this explanation. In this manner, radiation was ruled out as a cause. It was the "see no evil" approach to the protection of public health.

Similar PHS studies, also predisposed toward a viral cause, were conducted in the small Utah communities of Monticello, Parowan, Paragonah, and Pleasant Grove, where an unusually high number of leukemia cases also appeared in the 1960s. Again, no viral cause was found, and there was no independent assessment of the possibility of radiation being the cause. This was so, said Heath, who oversaw most of these investigations, because "to my way of thinking the burden of proof lies with the person who chooses to ignore them [the leukemia deaths] as merely chance events."[54]

The PHS investigators believed that the findings for all the clusters were inconclusive and that radioactive fallout was not the cause. In 1979 a panel of experts reexamined the PHS archives during a time of public upheaval on the subject and agreed with the first conclusion but not the second. They said that the documents did not demonstrate that fallout could be ruled out as a cause, which was a backward way of phrasing their finding.[55]

As a result of the mounting concern over iodine 131, the PHS took notice of the possibility of thyroid cancer existing among the downwind population in 1962. As with the leukemia and fallout link, the PHS also backed into this effort. Utah state health officials were advised by PHS investigators that "although the possibility of finding a relationship between any of the cases of thyroid cancer and the consumption of milk high in radioactive iodine content is most remote, we feel it necessary to study the situation both retrospectively and prospectively." Weiss and another PHS officer traveled to Utah and discussed the possibility of setting up a study.[56] With the heat on from Knapp and congressional hearings looming, even Gordon Dunning was urging the PHS to do something about iodine 131 in 1963.[57]

There the matter sat for two years until Weiss advised Gordon Dunning in June 1965 that the PHS was putting together a proposal for a thyroid study that would involve the examination of some 2,000 schoolchildren in southern Utah for "possible deleterious effects of fallout." Dunning also had a copy of Weiss's leukemia study in hand. He telephoned Weiss and told him that he favored the thyroid study, with certain modifications, because of the "compelling public relations need" for such a project.[58] But Dunning told his superiors something else. He foresaw "serious issues from the legal viewpoint" for the AEC and the possibility that claims and lawsuits might be filed against the AEC—as, indeed, was the case some fifteen years later. Dunning sent two communications to D. A. Ink, the AEC's assistant general manager, on August 27. The first stated, "Whereas it might not be wise to attempt to stop the PHS studies, perhaps we should consider the technical staff indicating more formally the fallacies upon which the studies are being based. Hopefully there might be a chance of averting a potential fallout scare by placing the purpose of the studies in proper context." The second memo was a variation on that theme. "Whereas, we cannot, or should not try to stop the PHS studies in Utah," it read, "it may be possible to make them realize the weaknesses in their justifications given for the study. Hopefully, this might influence them in their public expressions of need for the studies and also make them duly cautious in interpreting their results."

A third communication from Dunning to Ink several months later criticized the design of the study and voiced concern for "the welfare and happiness" of the children who might have to undergo biopsies. The AEC should be more forceful in its opposition to the thyroid study, said Dunning, who was pulling out all the stops. The memo ended: "I am not unmindful that we can rationalize that it is best not to appear to interfere with another agency and other reasons for not acting, but I cannot help but recall Abraham Lincoln's words, 'To sin by silence when they know they should protest makes cowards of men.'"[59]

The AEC scientist and his technical staff in the Division of Operational Safety, of which he was then the acting director, put together a detailed critique of the completed leukemia report and the proposed thyroid study. They recommended against publication of the leukemia report because, among other reasons, the phrase "excessive number" of leukemias was objectionable. To Dunning, the mildly worded Weiss study contained "unwarranted conclusions." Their objections to the thyroid study were based mainly on the assertion that doses had not been high enough to cause thyroid cancer.[60]

In early September, Peter Bing of President Lyndon B. Johnson's science advisor's office in the White House called Weiss's boss, Donald R. Chadwick, chief of the Division of Radiological Health, and asked, "What would be the federal government's liability for any clinical effects, possibly due to radiation, which might be discovered" during the thyroid examinations? It is not difficult to guess what agency fed Bing that question.[61]

After considerable discussion, it was decided to meet the next day—the promptness with which the meeting was called indicating its pressing importance. On hand for that meeting were a plethora of high-ranking bureaucrats: Bing, three HEW lawyers, Chadwick, Judson Hardy from the HEW information office, and Assistant Surgeon General Allen M. Pond. Representing the AEC were Ink, Dunning, and Dunham. Ink led off. He said that the AEC commissioners wanted certain changes in the press release announcing the thyroid study to take the onus off the current underground tests and place it on the atmospheric tests of the 1950s. The commissioners, Ink continued, "were quite concerned about the present version of the report on leukemia." Dunning then attacked the scientific validity of the leukemia report and the proposed methodology of the thyroid study. Bing was most concerned about communication between the two agencies. The note taker, Hardy of HEW, wrote for the record: "It became evident that the communication failure was within AEC, although this was not stated." The outcome of the meeting was a decision that any suits against the government would be handled on a case-by-case basis and that the AEC would suggest changes to the PHS in the press release and the leukemia report.[62]

Dunning pulled his usual sleight-of-hand trick after he looked at the draft PHS press release. He suggested to Ink:

I think it would be quite helpful if we could persuade PHS to drop the second paragraph of the announcement and replace it with the last paragraph of their announcement. It is correct that if we were pushed we would have to admit that the motivation behind this is fallout. On the other hand, rewriting the announcement per my suggestion (with some necessary minor changes throughout) could make it appear that the newer studies are a continuation of the already ongoing studies in Nevada and Utah.[63]

The necessary changes were made. The PHS had caved in. The surrender was silent and complete.[64]

Years later, during a brief period of mea culpa, F. Peter Libassi, the general counsel of HEW, explained the dereliction of duty of the public health agency thusly:

I think also it is important to note that the Atomic Energy Commission made positions available to the Public Health Service to hire persons to work in this area of health research, but it is also clear throughout the record that there was one agency in control at that time. That was the Atomic Energy Commission. I certainly don't believe that excuses any governmental agency for failing to reveal or make public the questions and doubts.[65]

Ink reported back to the commission at its September 10 meeting. The problems with both studies, he said, were "adverse public reaction, lawsuits, and jeopardizing the programs of the Nevada Test Site." The commissioners directed that a letter be sent to Surgeon General Luther L. Terry requesting that "all the broad experience in epidemiology" of the National Institutes of Health and the Communicable Disease Center be brought to bear on the leukemia report. This was akin to killing it with kindness. As for the thyroid study: "We do urge that the investigators exercise due caution in placing the study in proper perspective when presenting it to the public."[66]

Ink's letter was the last turn of the screw. The leukemia report disappeared from sight, for the time being. Weiss switched his efforts from leukemia to thyroid cancer. The PHS had in effect sacrificed one study to save another, but the thyroid study was to suffer a premature demise. So the agency wound up with very little to show the public for its efforts. The victor was the AEC, until it all began to unravel. The loser was the public.

During late September 1965 approximately 2,100 schoolchildren from the sixth through the twelfth grades in St. George and 1,450 from the same grades in Safford, Arizona, were visually examined for thyroid abnormalities. The PHS considered Safford, in southeastern Arizona, to be a community similar to St. George but one that had experienced little or no fallout. Since Safford also had a large Mormon population, it could serve as a comparison—or *control*, in scientific parlance.

The health service physicians who conducted the original screenings found that seventy of the St. George children and twenty-five of those in Safford had thyroid nodules or small lumps. The ninety-five children were then examined by a panel of nationally recognized thyroid experts. Of the seventy in Utah, twenty-eight were studied further, and of these, nine were selected for biopsies. None were recommended for biopsies in Safford.

A properly scrubbed PHS press release in late October downplayed fallout as a cause of the abnormalities. It stated, "The Surgeon General said it would be difficult to ascribe a definitive cause to any thyroid abnormalities found in the study since thyroid irregularities may occur naturally and information about the patterns of their occurrence is inadequate." In addition to checking the press releases with the AEC, the releases had to be cleared by the White House, where President Johnson was keeping track of the situation.[67]

The PHS was beyond its depth. It was dealing with national security matters when it was more accustomed to identifying different strains of flu. The fear was: What if cancer were found and the news media got hold of the story? Judson Hardy, public affairs officer for the Division of Radiological Health, advised the Utah state health officer, C. D. Carlyle Thompson, to discourage out-of-town reporters and photographers from going to St. George because they might disrupt the project. The PHS would not provide dose estimates to the press; this was up to the AEC, said Hardy. Chadwick put the dilemma this way to the surgeon general: "Exaggerated and unbalanced press accounts could hamper not only the Government's nuclear testing program but also peaceful applications of nuclear energy. On the other hand, any real or apparent attempt to 'cover up' the results of these studies would result in great damage to the reputation of the Federal Government."[68] Who, it might be asked, was looking out for the public's health at this time?

Chadwick need not have worried. The *Deseret News* story based on the PHS news release was headlined "Despite Nodules St. George Is Calm." The mayor, before departing on a deer hunting trip, said, "We're an easy going town and don't excite too much."[69]

But health officials continued to fret about uncontrolled press coverage. Near-panic and a flurry of memos ensued when Stephen Spencer, medical editor of the *Saturday Evening Post*, said he wanted to do a story on the thyroid tests. Hardy, citing "the sensitive nature of this entire subject," advised the agency to cooperate. He cautioned: "All Federal and State personnel interviewed [should] stay strictly within the limits of the existing public record and commonly accepted scientific knowledge." Over at the AEC, the public relations advice was: "the problem relates to atmospheric tests, not to our current underground testing program," which in itself was an interesting admission. There was, said Duncan Clark, director of the AEC's Division of Public Information, "no reasonable course of action other than that of cooperating with him."[70]

At the same time, Chadwick was outlining some of the very same sensitive concerns at an in-house critique of the thyroid study at the PHS research facility in Rockville, Maryland. He posed two questions: Are the cases related to fallout? and "What needs to be done to try to eliminate or rule out the possibility that fallout may be an etiological factor?" Chadwick was told that there were no findings as yet and that the study seemed to be statistically flawed. One point made at the meeting was that the study should be continued for another ten or fifteen years, since it was known that no thyroid cancers had begun to show up among the Marshall Islanders until nine years after exposure, and thyroid cancer, like other solid cancers, would very likely continue to occur until thirty-five or forty years after exposure. The meeting ended on an inconclusive note. Chadwick asked: "Assuming that we are not going to be able to pin this thing down etiologically, what should we be doing?"[71]

The PHS announced in March 1966 that no malignant tumors had been found among the children examined in St. George and Safford. The observed abnormalities were blamed on thyroiditis, an inflammation of the thyroid gland that is not caused by radiation. Residents of Utah, it was explained, had a history of thyroid diseases that was first noticed during draft examinations for World War I. *Time* magazine noted that Robert C. Pendleton of the University of Utah had dismissed the study as "the same old hunkum."[72]

Two months later a joint AEC-PHS meeting was convened to talk about how to proceed with the thyroid study. There was a wide range of opinions, as noted in the minutes:

> Some concern was expressed about the problem of public relations resulting from these studies. The problem was viewed differently by several persons at the meeting. Some would have us forget about the study; others to continue it; some were of the opinion that the study of an exposed population as now in progress should continue because of the valuable medical information that we are obtaining independently of the study of the possible effects of fallout; and others thought that to discontinue the study now may leave the PHS vulnerable to the criticism of "sweeping the problem under the carpet."[73]

Two reports of the "completed" study were published—one in the *American Journal of Medicine* by Dr. Marvin L. Rallison of the University of Utah, who had supervised the clinical and laboratory tests, and another by Edward Weiss in the *American Journal of Public Health*. The studies concluded that there were no differences in abnormalities between the children in Utah and those in Arizona, which is not surprising, since both authors were coauthors of the other's work and both had participated in the PHS study. Both recommended a longer-term study.[74]

What was interesting was not what the two published accounts contained but what they left out. In March 1970 Rallison, an assistant professor of pediatrics at the university's medical center, had written the acting state health director: "There appears to be a very disturbing trend characterized by the identification of an increased number of nodules among the children of Utah and Nevada who were potentially exposed to fallout radiation. If this trend is borne out by the re-examination by the panel in May, it could signify the appearance of thyroid damage as a result of the radiation fallout exposure about eighteen years ago." Apparently the panel did not support Rallison's tentative conclusion, because there was no further mention of it.[75]

Two years later Rallison wrote to a federal environmental researcher that he had "discovered a very discomforting observation which I had heretofore overlooked." In searching for possible causes of goiter in adolescents, "I compared the rates of adolescent goiter occurring in children exposed to fallout radiation so defined by residence history with those not exposed. I was surprised to find considerably higher levels among exposed children."[76]

William R. Bibb of the AEC's Division of Biology and Medicine drew the attention of a staff member of the Joint Committee on Atomic Energy to Weiss's comment that the short length of time was a limiting factor in the effectiveness of the thyroid study and predicted that it would be "a reasonable assumption" that the surgeon general would not pursue the study because of the low doses that had been estimated by the AEC. It was a correct assumption.[77] The Public Health Service ended the study in 1971, six years after it began, some eighteen years after the height of the fallout, and well before the end of the expected occurrence of thyroid cancer. The AEC was off the hook, for the time being.

When the fallout controversy was raised anew in the late 1970s, the PHS took another look at what had been done, or not done. The conclusion reached in 1981 was that the study was badly flawed and had ended too soon. The University of Utah was given a grant by the National Cancer Institute for a follow-up of the original study. Rallison was to reexamine 1,500 of the schoolchildren who had been examined in Utah and an equal number in Arizona.[78]

Federal health officials made one more attempt in the late 1970s to determine if there was an unusual number of leukemia deaths as a result of the atmospheric tests, this time concentrating on the soldiers who had witnessed Shot Smoky in 1957. There were differences between the basic situations of the military participants in the tests and the offsite civilians. The military was training for war. There was the potential for the soldiers to be exposed to radiation from the immediate blast as well as local radioactive fallout from the resulting cloud. The military lived for only a short time in the region, so ingestion of locally grown foodstuffs and milk was not a factor. They were lectured, shown films, told what precautions to take, and given film badges. The civilians were more likely to be exposed to repeated doses of radioactivity, about which they knew little or nothing.

The similarity was that the atomic veterans, as they came to be called, had done little better than the civilians in collecting compensation from the government. At the time of the Allen trial, of the 2,883 radiation claims that had been filed with the Veterans Administration, only sixteen were awarded as radiation related. The veterans were seemingly blocked from suing the government by the Freres Doctrine, a 1950 ruling by the Supreme Court that held that the government could not be sued for injuries incurred during military service. The ruling was being challenged in court, and there were legal attempts to circumvent it that were blocked by a bill that slipped through Congress with the support of the Reagan administration in 1984.[79]

The case of Paul Cooper, the veteran who died of leukemia in Salt Lake City, was referred to Glyn G. Caldwell, chief of the Cancer Branch of the Center for Disease Control. Caldwell launched a study in 1977 of the participants in Shot Smoky. Searching military records, Caldwell and his colleagues found a list of military personnel who held film badges, and they subsequently identified 3,244 men who had witnessed the shot. They then began the laborious process of tracking down the individuals. (It was Caldwell's publicized search that attracted the attention of Mrs. McKinney.) By March 1980, a total of 2,459 of the participants (75 percent) had been located and questioned. Nine cases of leukemia were found, where the expected number from national rates was 3.5.

From October 1980, when the preliminary findings were published in the *Journal of the American Medical Association*, to when
Caldwell testified at the Allen trial in 1982, two additional leukemia
cases were found among the 25 percent who had not previously been
questioned. Through questioning and mathematical computations,
four of the five causes of leukemia—genetic, viral, chemical, and
chance—were eliminated. What remained was radioactive fallout.
The levels of doses as recorded by the film badges of the nine victims initially identified ranged from zero to 3 rems, but Caldwell did
not think that the data on doses was reliable. The report concluded:
"The excess of leukemia cases among Smoky participants, therefore,
suggests either a greater dose than estimated or a greater effect per
rem at low-dose levels." This was an ominous finding that could also
be applied to the downwind population.[80]

If Knapp was the maverick within the AEC, Robert C. Pendleton
was the squeaky wheel among the otherwise well-greased scientists
in the academic community. During the 1950s, Pendleton worked at
the army's Dugway Proving Grounds in western Utah, where nerve
gas and other chemical weapons were tested, and at the AEC's Hanford facility before returning to the University of Utah in 1959 to
become the first fulltime director of the Department of Radiological Health. Pendleton stuck out in a number of ways. In a state of
teetotalers, he was a well-known amateur winemaker. An excellent
shot, he killed the deer and cougars whose carcasses he analyzed for
radioactive content. He also spoke his mind. Pendleton believed that
his first wife, who had died of cancer, had been a victim of fallout.
He tended to act on his personal beliefs, which was unusual for his
occupation.[81]

Through the 1960s and 1970s Pendleton, who was not shy of the
media, was a clear and persistent critic of the AEC, from whom
he also derived some of his research funds. For instance, when
the AEC announced that the radioactive fallout from a Plowshare
experiment in 1968 was infinitesimal, Pendleton replied in a Salt
Lake newspaper that this statement was "standard public relations
gobbledygook." He happened to be right on target. This was the

Schooner shot, whose fallout was detected in Canada. The Public Health Service, however, was incensed at Pendleton's comments. John R. McBride, the acting director of the test site monitoring effort, telephoned Utah state health officials to complain. Those officials said they needed more information "to be better prepared to develop news releases which will block adverse publicity such as happened in the last incident." The Utahans noted the "extremely cooperative attitude" of McBride. It was a love feast, with Pendleton on the outside.[82]

When federal funding for his laboratory dried up in the 1970s, Pendleton said, "They were getting pretty sick of me, I think." But others said the funding had ended because Pendleton could not organize himself sufficiently to publish results, to which Pendleton replied: "There are a lot of people who have opposed this [publication] for a variety of reasons who damned well don't want to see it come out because it's like the business of exposing their own culpability." Pendleton used the charge that he had been scuttled to make a pitch for more federal funds in the late 1970s. By this time university officials thought Pendleton was senile, and a colleague said he was ill, but he had his moments of lucidity. Pendleton recalled, "I'm the first person who claimed that the AEC was trying to make guinea pigs out of the people of Utah. I got into more difficulties as a result of that than I like to relate." Near the end of his life, Pendleton said he probably should have kept his mouth shut, "like almost all scientists do." He died shortly before the start of the Allen trial.[83]

Pendleton and iodine 131 met by accident. On the day following the Sedan Event in 1962, he was in the Wasatch Mountains above Salt Lake City with twenty students taking background readings of radioactivity. A reddish-brown dust cloud approached from the south across the Salt Lake Valley. When the cloud enveloped the students "our instruments went completely nuts and we couldn't measure what we were trying to do," said Pendleton. The readings were about one hundred times the natural background for the area, and it occurred to Pendleton that the high measurements might be from fallout. Next morning high readings were also obtained from lawns in the city.[84]

At the time, Pendleton was funded by the PHS for a study of the cessium 137 content of global fallout. But when Salt Lake and northern Utah were blanketed by local fallout from Sedan and other tests that July, the project was redirected to the radioiodine content of milk collected at sampling stations Pendleton had previously established. Milk samples were gathered and analyzed in the university lab. Pendleton notified the press and the Utah State Department of Health about the high iodine 131 content of the milk and suggested that it be processed into cheese, powdered milk, or condensed milk in order to lessen its radioactive content.

The health department was reluctant to cause economic harm to milk producers. State Health Director C. D. Carlyle Thompson complained that Pendleton's warnings and public statements "complicate the performance of our public responsibility and of our working relationship with industry." Without the confirmation of a hazard from the AEC or the PHS, the department delayed taking any action until after most of the contaminated milk had been consumed. Thompson then countered Pendleton's public statements by telling the newspapers repeatedly that there was no danger from the fallout. To Pendleton, the state health officials seemed too dependent on advice from the AEC.[85]

It was not surprising that Knapp and Pendleton began to correspond in the early 1960s, since they had radioiodine and Gordon Dunning in common. Dunning criticized a draft of Pendleton's paper that had been submitted to *Science* magazine and that was undergoing repeated reviews. The AEC scientist termed the report unscientific and biased, which was the ultimate insult. This raised Pendleton's ire.[86] He wrote Knapp:

> I suppose that it reveals a deficit in character to admit that the "fiery opposition" from Gordon Dunning is somewhat amusing. I say this because it is obvious that the opposition stems from a desperate attempt to coverup, and to avoid admitting that he was in error. Dr. Dunning's efficient machine-gunning of the provincials from Utah has not been forgotten here, and if he has to face some of this himself he might better appreciate the importance of knowing the target next time.[87]

At the time, the AEC was holding up publication of both the Knapp and Pendleton studies. Pendleton would have none of this. He wrote Knapp, "I do not intend to allow this extremely dangerous situation which has developed regarding the supra-legal position of the A.E.C. which can even control the publication of scientific articles in the national journals to go unchallenged." So the fiesty Utah scientist distributed three hundred copies on his own and was having an additional five hundred printed. The copies went to members of Congress, Utah state legislators, the governor, and newspaper editors. He added:

> I wonder if those in power in the Atomic Energy Commission realize that it is one thing to err and admit it, but that it doubles the guilt when you err, refuse to publish it and then deliberately go contrary to all the democratic principles by suppressing release of information which is vital to the nation.[88]

While Pendleton tended to shoot from the hip, Joseph L. Lyon of the University of Utah first approached the subject in the late 1970s with skepticism and annoyance. Lyon was aware of Pendleton's work while he was a student at the university's medical school in the 1960s. After getting his medical degree, Lyon obtained a master's of public health degree from Harvard University and returned to Utah, where he was an associate professor in the medical school. Lyon was a codirector of the Utah Cancer Registry. In his midforties, Lyon was typical of the new breed of scientist on campus—ambitious, analytical, and successful at obtaining federal funds for research. Lyon said, "Pendleton did not have much data, but he had a lot of fervor." They were two different breeds.[89] Lyon read Paul Jacob's article in the *Atlantic* in 1971. "Well, this seems to be somewhat more of the same overheated environmental hysteria," he thought. But the publicity about Paul Cooper and the atomic veterans caught his attention in 1977, as did Gordon Eliot White's front-page story in the *Deseret News* in August.[90]

The cancer registry began to get telephone calls. The callers were told that there did not seem to be any link between the tests and cancer, but at the same time Lyon realized that the records were complete only since 1966, when the registry had been established. Because the issue did not seem to be going to disappear and because one of Lyon's research assistants needed some work, a small study was designed to look into whether fallout caused any increase in cancer.[91]

Lyon doubted that the brief study would show any positive results. "I think if you had asked me at the time what the probability was of finding any kind of association, I would have answered there was about a 95% chance that there was no association," said Lyon. He based this assumption on the thyroid studies that had showed no link and the low radiation doses estimated by the government.[92]

In setting up the study, Lyon was well aware that the cancer situation was different in Utah from anywhere else in the nation. The previous year he had published an article in the *New England Journal of Medicine* that stated that deaths from all cancers were 22 percent lower in Utah than the national average. The state ranked fiftieth in the nation. The study noted that the Mormon church discouraged the use of tobacco, alcohol, tea, and coffee. Cancers related to tobacco and liquor were 55 percent below the national average.[93]

The problem was that, in order to obtain a true comparison, Mormons who had been subject to fallout had to be compared to other Mormons who had been subject to much less fallout or none at all. Lyon's solution, after studying fallout maps furnished by the Department of Energy, was to divide the state into two parts. The seventeen southernmost counties, all rural and containing about 15 percent of the state's population, were labeled the high-exposure area, and the remaining more urban, northern counties were designated the low-exposure area.

Leukemia was selected because of its short latency period and its known association with radioactive fallout. Children under the age of fifteen were chosen for study because they were the most susceptible to radiation-induced leukemia, thus increasing the chance of detecting an effect from fallout. The childhood deaths were divided into a high-exposure cohort, or group, and a low-exposure cohort. The high group lived in Utah during the period of atmospheric testing. The low group died before 1950 or were born after 1958. The time limits of the study were from 1944 to 1975. Death certificates were reviewed for this period and the residency and cause of death ascertained.

The preliminary results of the study emerged in early December 1977. The findings were positive and fit the time frame of the exposed group. Lyon was shocked. "My first reaction," he said, "was that we made an error somewhere, that we simply had to have made some mistake in either counting the deaths or in tallying up the population."[94]

The study was redone, this time verifying by hand what had previously been done by computer. In March 1978, Lyon concluded: "The differences did not appear to be in any errors we had made. They appeared to be the result of a phenomenon that was buried in the vital records of the State of Utah. There definitely was an excess of leukemia during that period of time."[95]

It would be another six months before the first claims were filed by Haralson, but the public controversy was already heating up. Neither Lyon nor his coworkers were of Pendleton's ilk. The public spotlight and all that it entailed, and how they might be viewed by fellow scientists and those who handed out federal grants, greatly concerned them. It was readily apparent that publication of the study would involve the researchers in a major controversy in the public arena. The question at this decisive point was what to do with the results. Lyon and his coworkers debated whether to become involved or not. "We had several discussions among ourselves as to whether we quietly filed it away and went on to something more productive, or whether we published," said Lyon. They finally decided that there was no ethical alternative but to make the results known, so they readied the study for publication in the *New England Journal of Medicine*, which had strict rules against prepublication news stories on its articles.[96]

Word of the results of the study leaked out, and Lyon received phone calls from newspaper reporters and lawyers. He gave out no details. If he had, said Lyon, it would have been "at the peril of my own life and career." However, he briefed Department of Energy officials in Washington and applied for federal funds to expand the study.[97]

In mid-December, Lyon attended a meeting on fallout at the

Utah State Department of Health and was given a packet of information that had been gleaned from the state archives. He did not
open it until a few days later. Lyon was unaware of Weiss's unpublished leukemia study and was greatly surprised to find it among the
other materials in the packet. He thought he had done a careful literature search. Although the Weiss and Lyon studies were designed
differently, the basic data matched. That was reassuring to Lyon.

Of Weiss's conclusion that the only causal link between excess
leukemia deaths and fallout was residency, Lyon said, "I thought
that was a remarkable statement. It was, to me, akin to someone
saying that other than being a lifetime smoker, there was nothing to
connect a person's death from lung cancer to cigarettes. It seemed
quite remarkable that one could look at an area where you had some
evidence of exposure and somehow completely ignore it and assume
something else must have been going on." Lyon thought the study
should have been published.[98]

At this same December meeting the University of Utah stepped
up its extremely aggressive pursuit of federal funds for health studies.
The meeting of the State Fallout Steering Committee was chaired by
Chase N. Peterson, the university's newly arrived vice-president for
health services. Besides state and university health officials, Caldwell was there from the Center for Disease Control. He voiced some
cautions and posed some questions in a neutral tone. A naive Peterson had warm praise for Caldwell. He said that although Utahans
distrusted outsiders and the federal government, Caldwell was an
exception because of the enthusiasm he had voiced for Utah's bid.[99]

Caldwell had other thoughts on his mind. He wanted to undertake
the study that the university had in mind, but the university controlled the data. When he returned to Atlanta, Caldwell expressed
some grave reservations about the Utahans. He wrote a memo for
the record (a favorite PHS device) stating that the university was
going to seek "large studies" and Lyon was "going to push very hard
to be the sole supplier" of such studies. "I see the federal role being
extremely limited by the state by limiting our access to data," said
a disappointed Caldwell. Others would come up against the same
roadblock.[100]

The federal health establishment, too, had mechanisms for delay and denial. In November, Lyon had submitted a draft of the forth-coming article that he was so jealously guarding from prying out-siders to the National Cancer Institute as part of a pitch for a five-year, $3 million follow-up study. The university was the ideal institution to undertake such a study, said Lyon, because it had no vested interest in either positive or negative outcomes and was "in-sulated via carefully constructed mechanisms from direct political pressures." [101]

The institute's staff did not think the study was a good idea, and it was quickly picked to pieces by the federal health establishment. The cost seemed excessive, and staff members, including Charles E. Land, felt that Lyon's interpretation of the data was open to ques-tion. The study seemed to be of social and political interest rather than of scientific value and thus should not be funded with monies earmarked for scientific purposes. Nor did it seem likely that new scientific knowledge would emerge from such a study. [102]

A tristate proposal involving Utah, Nevada, and Arizona had been shot down in November 1979, and an $18 million study spearheaded by the University of Utah was disapproved in September 1980. There was anger and frustration in Utah, where the feeling was that the state was being stonewalled by reluctant bureaucrats. Lyon believed that Peterson was inept at obtaining grants. "Chase thought that he had the goose that lays the golden eggs," said Lyon. [103]

There was barely concealed contempt in Washington. "The strat-egy posed by the Utah investigators is potentially wasteful of large resources which are nowhere justified in the proposal," wrote the reviewers, who included Clark Heath and others from the federal health establishment. [104]

The Lyon study was published in the February 22, 1979, issue of the *New England Journal of Medicine*. The statistical implications of the study were clear. The chance of a child dying from radiation-induced leukemia was greater the closer the youngster lived to the test site, and, it could be inferred, the chance for an adult was some-what the same. The study was the most thorough to date on the link between cancer—in this case childhood leukemia—and the offsite population. It was bound to be controversial, and it was. [105]

The numbers, as in the other studies, were rather small. Among

the high-exposure cohort in the northern part of the state, 152 leukemia deaths were found where 119 would be expected. For the high-
exposure cohort in the seventeen southern counties, where there
were far fewer people, there were 32 deaths where 13 would be
expected—2.4 times the normal level. From the seventeen southern counties, Lyon selected the five closest to the Nevada Test Site
and found a level 3.4 times normal. There was a significant excess of
deaths in southern Utah between 1959 and 1967, a time frame that fit
the expected peak period for leukemia in relation to the atmospheric
tests.

The biggest problem Lyon encountered in putting the study
together was the lack of adequate information on doses. He believed that the AEC's 1959 estimates were unreliable and underestimated.[106] By taking the excess number of deaths and estimating the
dosage necessary to produce them, Lyon said it would have taken
a dose of between 4 and 10 rads to the bone marrow to cause that
number of leukemia deaths.

Lyon was cautious in his written conclusion on the cause. "The
increase in leukemia deaths could be due to fallout or to some other
unexplained factor," he wrote. Lyon was more emphatic in his congressional testimony and at the trial, where he said, "The most likely
cause is radiation. The size of the difference and the geographical
location of it all made me come to that conclusion."[107]

Lyon's study and testimony were central to the plaintiffs' case on
causation. He testified that a child who lived in the seventeen southern counties during the period of atmospheric testing and who later
died of leukemia had a 59 percent chance of the leukemia having
been caused by the fallout. For a similar death in the five counties
closest to the test site, the chance was 71 percent. The law did not
require absolute certainty, only a greater probability than not, which
was what Lyon seemingly furnished for the leukemia victims.[108]

Lyon had his critics, and chief among them was Charles Land of
the National Cancer Institute. Land was a health statistician, having
received his doctor's degree in statistics from the University of Chicago. He worked for the government-funded Atomic Bomb Casualty Commission, which studied the effects of the two bombs that

had been dropped on Japan, and joined the institute in 1975. Land helped prepare the controversial BEIR III report and was a member of those organizations that set the radiation standards that Morgan was so critical of. Land and Lyon were approximately the same age, but one was inside the government and the other was outside.[109]

At the invitation of the editor of the *New England Journal of Medicine*, Land wrote an editorial comment on Lyon's article that appeared in the same issue of the magazine under the caustic headline "The Hazards of Fallout or of Epidemiologic Research?" Land suggested that the Utah doctor should have selected a control group from outside the state and added: "It is in the nature of things for cancer mortality rates based on small populations to vary widely over time." He said there were pitfalls in such a study, and he damned it with faint praise at the trial.[110] "Generally speaking," he said, "I thought that the study was not a bad idea. I liked in particular some of the logic that went into it. I thought, however, that it was a marginal finding and that not much could be concluded from it and that it should be treated very cautiously."[111]

An independent assessment of Lyon's work and Land's criticism was made by Herman Chernoff, a statistician in the Department of Mathematics at the Massachusetts Institute of Technology. In a critique of both sides of the argument done for the General Accounting Office, the investigatory arm of Congress, Chernoff wrote: "In short, the Lyon et al. paper is basically sound. Although they do not claim it, I feel that fallout probably caused an increase in leukemia. Proof does not and may never exist, and Land is correct in stating that the results of the study should be interpreted with caution and that it would be desirable to have certain types of additional data."[112]

After publication of the Lyon study, the Utahans again took up their unseemly pursuit of federal research monies. Peterson, stating that "I'm obviously making a pitch in the baldest of terms," filled in members of Congress on the significance of the proposed studies in the most effusive terms. The University of Utah official said, "I honestly think it ranks with discovery of the germ theory of medicine or the discovery of penicillin in terms of how we're going to live with ourselves and our civilization over the next century."[113]

In late September 1980, Secretary of Health and Human Services Patricia Harris announced that five federal agencies, including the Department of Energy and the Department of Defense, would make $4 million available to scientists in Utah, Nevada, and Arizona to study the health effects of the fallout. Negotiations between the parties were to be undertaken by Donald S. Fredrickson, director of the National Institutes of Health.[114] But nothing happened. Peterson guessed that none of the federal agencies had ever intended to act on this matter. Meanwhile, Peterson, who would shortly become president of the University of Utah was having problems closer to home.[115]

It had become evident that, rather than a straight grant to Utah, the process would have to be competitive. The University of California at Los Angeles had emerged as the main competitor. This worried Lyon, who told the governor's press secretary that the UCLA bid was "worrisome because this particular researcher has been quite vocal all along in insisting there is no correlation between the open air tests and ill health effects." So much for the "no vested interest" argument made earlier.[116]

But Lyon need not have fretted. The UCLA researcher would need the Utah data that was controlled by Peterson, who neatly sidestepped this dilemma. Peterson said that there had been a prior problem with UCLA and that "We simply cannot be sure that confidentiality will be protected even under the best of circumstances with an outside or unrelated researcher."[117] He then advised staff members of the Senate Committee on Labor and Human Resources:

> As a final caveat, it should be emphasized that access to the data base within the Utah Cancer Registry and the L.D.S. Church is absolutely essential to perform appropriate studies. I bear responsibility for reviewing applications for access to these data. In spite of a written request to at least one of our contract competitors [UCLA] and numerous telephone requests, no one at the University of Utah has ever received an official request for use of these data. By definition, this should render competitors ineligible to perform the studies.[118]

As Lyon later put it: "There is no competition when the university holds a lock on the data."[119]

Meanwhile, Senator Orrin Hatch, who was chairman of the relevant Senate committee and had considerable influence within the Reagan administration, had to be brought in on the matter by the Democratic governor. The two were not friends, and Maggi Wilde, an aide to Matheson, said that it would be distasteful to enlist Hatch's help but that it would be "the smart way to proceed" because the senator "obviously chairs the right committee." Some of the limelight might have to be shared, said Wilde, who added that it would be best to keep this memo in-house because of its candor.[120]

Hatch and Matheson went to work on the Reagan administration, and in May 1982 Secretary of Health and Human Services Richard S. Schweiker advised the senator that a five-year contract for $6.4 million would be awarded to the University of Utah. There were three parts to the study. Lyon, who was the principal investigator, would do an expanded epidemiological study of leukemia and thyroid cancers. Rallison would follow up on the PHS thyroid investigation, which had been prematurely ended, and McDonald E. Wrenn, director of the Division of Radiology, and Charles W. Mays, who had been Pendleton's colleague in the early 1960s, would attempt to determine doses from radioactivity in building bricks. Fifteen to twenty researchers would work on the various projects. Peterson said, "We're looking for the truth."[121]

The 1979 Lyon study was the last that was basically in place before the subject became polarized by the adversary system of justice. Other studies were presented at the trial, and their results, not surprisingly, fit the needs of those who commissioned them, one having been done for the government and the other for the plaintiffs. Both studies were later formalized when they appeared in scientific journals. It remained to be seen whether the university's studies would be completed in time and be definitive enough to affect anything.[122]

During two months of testimony that was unrelenting in its somberness, Indian summer and fall had passed. It was now winter. The Allen trial ended on November 16. Snow blanketed the surrounding mountains when the judge and the eleven attorneys returned to the courtroom a month later for closing arguments. The oak-paneled courtroom in the granite courthouse was packed with members of the press, who found the summation an easier event to cover than the many days of plodding testimony. All the nation's major print and electronic media were represented, along with some foreign journalists. Japanese journalists had visited the trial; it held special relevance for their readers and viewers.

Ralph Hunsaker led off for the plaintiffs. Clad in a dark blue three-piece suit, a more formal attire than he had customarily worn in court, the Phoenix attorney began:

> May it please the court and counsel. This case, I believe, is a
> case of really first impression before the judicial system of the
> United States. It is of great magnitude. I think it presents to
> this court for decision some rather significant social and legal
> issues and an opportunity, I believe, for this court to cure a
> rather grave wrong which has occurred in the history of this
> nation.[123]

Dale Haralson followed with a description of the damages. The Tucson lawyer sipped water continuously to sooth his irradiated throat. His left shoulder drooped, and his clothes were too large for his shrunken frame. Normally a short man, Haralson seemed even smaller. But he would survive. Haralson began softly:

> You know, your honor, whenever I get up to talk about the
> value of human life, I feel inadequate. For four years and seven
> months I've waited for this opportunity. As I reviewed my ma-
> terials, I've never felt more inadequate to address this question
> —to talk about the value of human life that was taken as a result
> of the greatest civilian tragedy that has ever been experienced
> by United States civilians since the Civil War.[124]

After the noon recess, the government presented its closing arguments. Henry Gill paced about the lectern, the very picture of a
Washington lawyer in a dark blue pin-striped suit. He said:

The issues were proximate cause, negligence, damages—as in any tort case—as well as statute of limitations and discretionary function. The burden of proof on each of these issues was in accordance with horn-book law and in accordance with your decision was laid squarely on the shoulders of the plaintiffs. And now the task is for the court to decide whether the plaintiffs have met the burden as to each and every one of these issues; based on the record, not on the rhetoric.

Gill continued:

This case is in that record. It's not in the newspapers. It's not in the congressional records. It's no place but in the record that's before the court. I spent many, many hours with that record, as did all of the people sitting at this table, and the one thing that comes out so strikingly and so glaringly to me, all the other issues aside, is the extreme weakness of the plaintiffs' causation case and the overwhelming strength of the government's causation case. Indeed, the clear and convincing evidence is that the cancers and leukemias of the plaintiffs in this case were not caused by the activity of the government at the Nevada Test Site.[125]

After listening to the rebuttals, Judge Jenkins said, "As I was sitting here, I remembered a line from *My Fair Lady*: 'I've grown accustomed to your face.' I appreciate your being here. I'm rather going to miss you. I'll do the best that I can within the time that it takes to turn out a determination on the matter."[126]

CHAPTER 9

THE JUDGMENTS

THERE WAS a basic unfairness
in the situation. A single judge had to decide what Congress and the
executive branch of government, with their vast resources, had failed
to act upon. The judge had a massive trial record filled with technical
complexities and discrepancies to consider and one law clerk to help
him. The record consisted of 6,825 pages of transcript containing
the testimony of ninety-eight witnesses and the intermittent pleadings of eleven lawyers. Nearly 1,700 documents, totaling 54,000
pages, were submitted as exhibits. These exhibits were stored in
nineteen cardboard boxes, and pretrial depositions required another
five boxes. There were other extraneous materials that the judge also
considered.

Judge Jenkins handed down his 489-page opinion on May 10,
1984, seventeen months after the closing arguments. Ten of the
plaintiffs were awarded damages. For the remaining fourteen test
cases, the proof of causation was not sufficient. Even with the split
decision, the judgment was hailed as a triumph for the plaintiffs. It
was the first time that a federal court had determined that nuclear
tests had been the cause of cancers.[1]

"I've never been so elated in my life," said Dale Haralson. Government lawyers refused to comment. Walter Sullivan wrote in the
New York Times that the judgment was a "landmark ruling." The
Jenkins decision, it was said, might bear on suits filed by other victims—the sheep ranchers, test site workers, Navajo uranium miners,

and servicemen. Yet, for all this, as Mrs. Lorna Bruhn of St. George was quoted as saying, "How can anyone call it a victory? All the money in the world won't pay for what we lost."[2]

The judge decided against the government on the statute of limitation and the discretionary function exemptions. Vague suspicions of a link between the tests and cancer did not count as the starting point for the two-year period. Hard information was needed, and that was lacking for the victims and their relatives. The decision to conduct the tests and their specific design were within the realm of policy, but what happened after detonation was an operational matter and thus was not covered by the discretionary function exemption of the Federal Tort Claims Act.[3]

Judge Jenkins cited a hypothetical situation to illustrate the non-policy aspect of the discretionary function question:

> Suppose a high-level decision maker says, "International pressures make open-air atomic testing highly necessary. Time is of the essence. We cannot tell our own people. We just need to do it and do it as fast as we can. We know as a result of such testing some people are going to get hurt. We can't tell them they are going to get hurt. We can't even warn them what to do to minimize or prevent the hurt. In order to preserve our way of life some people unknown to them and unknown to us are going to give their all for the good of all."

That was a policy decision and thus was exempt from the tort claims act. But that was not the case in this situation, said Jenkins (although in the aggregate, it certainly seemed to be). He wrote: "Nowhere in the 7,000 pages of the record was that kind of a decision made, eyes open, as a high-level or low-level policy decision. Nowhere were there to be human guinea pigs as a matter of policy. That is simply not here."[4]

The government's offsite monitoring program was negligent. Internal doses were not measured, and external gamma dose estimates "amounted to, at best, an educated guess." Government estimates were minimum figures, and it was "extremely likely that many people received more exposure to alpha, beta, and gamma radiation both externally and internally than more conservative guesses have indicated."[5]

The safety precautions taken for workers at the test site and in the national laboratories were not available to the offsite population. "In circumstances in which test site personnel have reason to predict that exposure may approach—or exceed—the 3.9 r standard, the scientific justification for monitoring workers directly, but not the people around them, especially children, defies the imagination," wrote Jenkins.[6]

The public information program was "woefully deficient" in three ways: offsite residents were not told of cancer risks, they were not instructed in simple precautionary measures that could have protected them, and the warnings to go indoors "failed to provide enough information soon enough to be useful and effective." Public statements by the AEC did not warn or educate; they just reassured.[7] Jenkins added that the statements "demonstrate that responsible persons at the operational level of continental nuclear testing neglected an important, basic idea; *there is just nothing wrong with telling American people the truth*" (emphasis is Jenkins's).[8]

The judge then dealt with causation. He wrote:

> In the law, as in science, one always faces uncertainty. This court, faced with the duty of judgment in this case, does not have the luxury of the zealous absolutists who "know beyond doubt" that each and every cancer in the Great Basin is the result of open air atomic testing, or of their absolutist counterparts who "know beyond doubt" that none resulted. The court is disciplined by the record and the application of rules of law.[9]

The judge used three primary and five secondary factors to determine causation in each specific case. The primary factors were: the probability that a plaintiff was exposed to fallout, the fact that the cancer was of a type known to be caused by radiation, and residency in the downwind area between 1951 and 1962. If all the primary conditions were satisfied, then he considered the secondary ones, which were: extent of exposure, sensitivity factors such as age and affected organ, dose estimates, epidemiological studies, and an appropriate latency period.[10]

Using the above guidelines, Judge Jenkins awarded damages to the eight leukemia victims, four of whom were children. They were *Karlene Hafen, Sheldon Nisson, Peggy Orton,* and *Sybil Johnson.* The adults were *Arthur Bruhn, John Crabtree, Lavier Tait,* and *Lenn McKinney.* Lyon's testimony was decisive in these cases. *Jacqueline Sanders,* who had thyroid cancer, was also a successful plaintiff, as was *Norma Pollitt,* who had breast cancer. Jenkins noted that the thyroid study conducted by the Public Health Service had ended too soon.

The monetary awards were modest. The total damages awarded to the nine successful plaintiffs was $2,660,000 (Mrs. Pollitt's damages were not fixed because she had died between the end of the trial and the judgment). For the relatives of the eight leukemia victims, a spouse averaged $240,000, a son or daughter $40,000, and a single parent was awarded $250,000, with a couple splitting that amount. Mrs. Sanders was awarded $100,000.

Despite the massive record, the sum total of knowledge that Judge Jenkins could bring to bear on the unsuccessful plaintiffs seemed quite fragile, although more was supposedly known about the effects of radiation than any other toxic substance.

When *Willard Bowler* had died, he had skin cancer. The increase of skin cancer among the offsite population was insignificant. Because he was outside most of the time, Jenkins reasoned that natural radiation from the sun may have been the cause.

Melvin Orton had died of stomach cancer, for which the plaintiffs' study found no increase in men among the downwind population. The study, which had been specially prepared for the trial, backfired in a number of cases.

The kidney cancer of *William Swapp* and the prostate cancer of *Lionel Walker* were of a type for which no relationship with radiation had yet been established.

Kent Whipple and *Delsa Bradshaw* had died of lung cancer. The plaintiffs' study found no increase in lung cancer. However, the Japanese data indicated that an increase in lung cancer might show up at a later date, the judge noted.

Geraldine Thompson and *Donna Berry* had died of ovarian cancer, for which there was little evidence of increased incidence. The ovaries and uterus have a lower sensitivity to radiation than other organs.

The sudden death of *Lisa Pectol* was unofficially attributed to a brain tumor, which can be caused by radiation. But no autopsy had been performed, which would have confirmed the cause of death.

Catherine Wood had died of cancer of the colon, for which a statistical link was sketchy. Also, her doctor, a cancer specialist, had discounted radiation as a cause of death.

Ten months before her death *Irma Wilson* had been diagnosed as having cancer of the bladder, but the actual cause of her death was uncertain. The bladder has a low susceptibility to radiation, and there was no evidence of an increased incidence of such cancers among the offsite population.

Daisy Prince had died of lymphoma. The link between radiation and lymphoma, in her case, was not convincing.

Glen Hunt had died of cancer of the pancreas, which can be caused by radiation but is more often caused by smoking. Hunt smoked up to two packs of cigarettes a day and drank coffee.

Jeffrey Bradshaw suffered from Hodgkin's disease, for which the plaintiffs' study found no increase among those surveyed.[11]

The judgment, in what *Newsweek* magazine called "the litigation of the century," was thorough in its citations, convincing in its reasoning, and wise in its conclusions. However, it was also limited in its scope. The opinion did not provide a clear guide for processing the more than a thousand remaining claims, and it lacked a ringing quality that the subject seemed to cry out for. There was a certain thinness to it, a lack of resonance that, had it been present, would have imprinted the matter clearly on the public consciousness and given it the permanence that it deserved.[12]

The judge was constrained by the limitations of the law. The law was not concerned with literary values or personal feelings. Nor did the law touch upon the range of human behavior that history had raised from the discordant muck of the past. Imbedded in this tragedy, this very American nuclear tragedy, were vivid examples of incompetence, arrogance, deception, avarice, deceit, treachery, and fraud. The responsibility for the betrayal of these citizens ranged from the presidents of this country to the bureaucrats they nominally controlled, from Congress to the judiciary, and to a lesser extent, from the academic community to the media. The law was also blind to morality and compassion, and the law was not properly vengeful.

The ultimate test of this judgment, one that the lawyers and the judge were well aware of throughout the proceedings, was that it had to stand up under the scrutiny of an appeals court, and that was exactly where the federal government promptly took it. The government could have ended its battle against its loyal citizens at the trial court level and retreated from the field with partial honors. Instead, with a thoroughness, a seeming vindictiveness, and a lack of feeling—all of which were consonant with the historical record—it pursued the matter to the court of last resort.

First, the sheepmen were finally defeated. A three-judge panel for the Tenth Circuit Court of Appeals in Denver ruled that no evidence had been withheld from the plaintiffs, government lawyers had not given misleading answers to interrogatories, and no veterinarians had been pressured to change their testimony. The appeals court concluded that "everything was then available to plaintiffs and they made their choice as to what to do and what to use." Furthermore, no fraud had been committed upon the court. Fraud occurred when a judge was corrupted or "influenced," or did not perform his judicial function, which was not the case in this matter. The panel called the plaintiffs' claims of fraud extravagant and reversed Judge Christensen's ruling.

It was a stinging rebuke to the trial judge who had heard both sheep cases. Said attorney Dan Bushnell, "Seldom has a panel in this Circuit so bluntly reversed a trial judge, and seldom has this Court ventured so far beyond its normal appellate powers." The ranchers fared no better in a rehearing by all the justices of the appeals court, which was known for its extreme pro-government stance.[13]

The appeals court reversal was sustained by the United States Supreme Court in January 1986 when, by a 5–3 vote, it refused to hear the ranchers' appeal from the appellate court ruling. Chief Justice Burger disqualified himself from the case, apparently because of his association with it as assistant attorney general some thirty years earlier. Burger retired from the court shortly thereafter and assumed the duties of chairman of the Constitution's bicentennial celebration. McRae N. Bulloch, a Cedar City rancher whose father was a plaintiff in the original suit, termed the Supreme Court decision "a great disappointment, but a kind of relief. We won't have our hopes up anymore."[14]

Once again, the sheep were miner's canaries for the humans. Two years later the tenth circuit court reversed Jenkins's judgment. The Denver appeals court based its opinion on the Allen case solely on the discretionary function exception to the tort claims act. It pointed out that throughout the thirty-five-year life of the act the federal courts had wrestled with this clause. Where before it had been construed narrowly, now the exception was to be viewed as more encompassing. The rules of the game were radically changed in midcourse for the Allen plaintiffs. It was but one more injustice for the victims of the system.

The conservative-minded justices followed an argument of judicial restraint and cited as the pivotal opinion a 1984 Supreme Court ruling handed down after the Jenkins opinion. In *United States* v. *Varig Airlines*, said the three-judge panel, the Supreme Court had rejected the distinction between policy-level and administrative-level decisions. Regardless of the level, according to the Supreme Court, "Where there is room for policy judgment and decision there is discretion." This view of the exception to the act shattered the plaintiffs' contention that there is a difference between actions taken at the executive-administrative level and those taken at the operational level, which previously had been the dominant Supreme Court interpretation. "It was irrelevant to the discretion issue whether the

AEC or its employees were negligent in failing to adequately protect the public," wrote the judges, who went on to state: "Plaintiff's entire case rests on the fact that the government could have made better plans. This is probably correct, but it is insufficient for FTCA liability."[15] The opinion pointed out that the federal government, had it been a private party, could not have escaped the responsibility for the injuries and deaths on the basis of sovereign immunity. The court expressed sympathy for the cancer victims but noted that there were "administrative and legislative remedies." The justices did not point out that these types of remedies had already been sought in vain and that what they were advocating was a Catch-22 situation.[16]

Judge Jenkins's opinion was reversed in these terms:

> The bomb testing decisions made by the President, the AEC, and all those to whom they were authorized to delegate authority in the 1950s and 1960s, were among the most significant and controversial choices made during that period. The government deliberations prior to these decisions expressly balanced public safety against what was felt to be a national necessity, in light of national and international security. However erroneous or misguided these deliberations may seem today, it is not the place of the judicial branch to now question them.[17]

In January 1988 the Supreme Court refused to hear an appeal from the appellate decision, and at that point the legal challenge ended.[18] But the question posed by the appeals court remained: Whose responsibility was it to question those faulty decisions and provide a remedy?

Sovereign immunity had prevailed, but at a terrible cost. In these times, given this case study and other examples of malfeasance, the government seemed more intent on pursuing its hidden policies than on benefiting its citizens. The disease of deceit was the most verifiable malignancy in these cases, and very few came away from this story without being contaminated.

NOTES

THE FOLLOWING abbreviations are used in the notes:

TT	Trial transcript from the Allen trial
EV	Material submitted in evidence at the Allen trial
DP	Deposition for the Allen trial
CIC	Coordination and Information Center, Las Vegas
JH, vols. 1, 2, 3	Printed record of the Joint Hearings of the House Subcommittee on Oversight and Investigations, the Senate Health and Scientific Research Subcommittee, and the Senate Committee on the Judiciary, 1979.

Volume 1 of the joint hearings is Serial No. 96-41, volume 2 is Serial No. 96-42, and volume 3 is Serial No. 96-129. Although Serial No. 96-129 has no volume designation, I employ a designation for ease of reference.

Prologue: The Crime

1. *Albuquerque Tribune*, May 18, 1953. The story was from United Press, a national wire service, so it is safe to assume that more than one newspaper ran it.

2. Department of Energy (DOE), "Announced United States Nuclear Tests," Office of Public Affairs, Nevada Operations Office, January 1983, p. 4; Atomic Energy Commission (AEC), minutes of meeting, May 21, 1953; JH, vol. 1, p. 142.

3. John S. Malik, TT, p. 2025; AEC/Department of Defense (DOD), Test Information Office, press release no. 48, April 30, 1953.

4. Defense Nuclear Agency (DNA), "Shots Encore to Climax: The Final Four Tests of the Upshot-Knothole Series, 8 May–4 June 1953," DNA 6018F, 1980, pp. 82, 90; AEC/DOD, Test Information Office, press release no. 65,

May 15, 1953; AEC/DOD, Test Information Office press release no. 66, May 19, 1953.

5. National Oceanic and Atmospheric Administration (NOAA), "Analysis of Upshot-Knothole 9 (Harry) Radiological and Meterological Data, Weather Service Nuclear Support Service," April 1981, p. 11, figs. 9, 10; Malik, TT, p. 2030; extraneous notes in "Journal, Shot IX," Fall-Out Plots, 0200, 0500, and Fall-Out Forecast From Briefing, winds valid 0505. These handwritten notes appear to have been made by the offsite radiation officer or an aide at the test site.

6. Testimony of Alvin C. Graves, Test Director, in *David C. Bulloch et al.* v. *United States of America* (the first sheep trial), 1956, JH, vol. 1, pp. 758–760, 829.

7. Defense Nuclear Agency photographs in author's collection, cleared for public release on August 15, 1979, and August 8, 1981.

8. Samuel Glasstone and Philip J. Dolan, eds., *The Effects of Nuclear Weapons* (Washington, D.C.: Department of Defense and Energy Research and Development Administration, 1977), pp. 27–32, 37.

9. Department of Defense, "Compilation of Local Fallout Data from Test Detonations 1945–1962 Extracted from DASA 1251; Vol. 1, Continental U.S. Tests," Defense Nuclear Agency, May 1979, vol. 1, pp. 149–154; Glasstone and Dolan, *Effects of Nuclear Weapons*, p. 29.

10: Communications logs for Shot Harry, EV, defendant's exhibits nos. 2–9. The material varies in completeness from the full messages to a synopsis of them.

11. Ibid., entry for 0415.

12. Ibid., message from Johnson to G-50.

13. Testimony of Dan Sheahan, *Bulloch*, JH, vol. 1, p. 1255.

14. Ibid., pp. 1165, 1254, 1143, 1146–1147, 1156–1158, 1184.

15. Ibid., pp. 1149–1150.

16. Sheahan in *Bulloch*, pp. 1196–1197.

17. Communications logs, May 19, 1953.

18. Ainslee Sharp, interview with author, Alamo, Nev., May 3, 1983; handwritten note attached to communications logs, entitled "Road Block-Alamo-Smithson," p. 77.

19. AEC/DOD, Test Information Office, press releases nos. 66, 67, 68, May 19, 1953.

20. AEC/DOD, Test Information Office, press release no. 70, May 19, 1953.

21. Louise Whipple Aicher, DP, July 27, 1982, pp. 42–45, and DP, September 22, 1980, pp. 15, 29–30.

22. Aicher, DP, 1980, p. 18; Keith Whipple interview with author, Hiko, Nev., May 3, 1983. Then there was the story of the "ghost of Alamo." An

AEC monitor pulled up in his vehicle, took a radiation reading, stated, "It's too hot here for me," and drove off. Department of Energy, "Proceedings of the Offsite Monitors Workshop," Nevada Operations Office, vol. 2, June 1980, p. 127.

23. Aicher, DP, 1980, pp. 20–21; Aicher, DP, 1982, pp. 16–17, 20; Howard L. Andrews, TT, pp. 4489, 4491, 4493; Office of Test Site Organization, press release, September 27, 1958, EV, defendant's exhibit no. 62B.

24. Aicher, DP, 1980, pp. 23–24; Keith Whipple interview.

25. Glasstone and Dolan, *Effects of Nuclear Weapons*, pp. 583–586, 594–597; Committee for the Compilation of Materials Caused by the Atomic Bombs in Hiroshima and Nagasaki, *Hiroshima and Nagasaki: The Physical, Medical, and Social Effects of the Atomic Bombings* (New York: Basic Books, 1981), pp. 130–140; John W. Gofman, TT, p. 3475; Graves in *Bulloch*, JH, vol. 1, p. 768; K. Z. Morgan, TT, pp. 2811–2813. The figures vary somewhat, but they are all in the same ballpark. I have chosen to use those cited by Morgan, an eminent health physicist, at the Allen trial. In many ways, however, the 1956 testimony of Graves is the most interesting because he had been exposed to a great deal of radiation, was the scientist in charge of the tests, and represented AEC thinking at the time. Graves said that at 25 roentgens there would be some change in the blood, at 75 to 100 roentgens nausea would occur, at 175 to 200 roentgens hair would be lost, and at 450 roentgens half of those exposed would die. Graves said the lower limits of a fatal dose were not known.

26. Committee on the Biological Effects of Ionizing Radiation, *The Effects on Populations of Exposures to Low Levels of Ionizing Radiation: 1980* (Washington, D.C.: National Academy Press, 1980), pp. 2–3, 30, 137; Glasstone and Dolan, *Effects of Nuclear Weapons*, p. 592; John W. Gofman, *Radiation and Human Health: A Comprehensive Investigation of the Evidence Relating Low-Level Radiation to Cancer and Other Diseases* (San Francisco: Sierra Club Books, 1981), p. 305. This simple sentence masks the intense scientific debate epitomized by the cited works.

27. John Cairns, *Cancer: Science and Society* (San Francisco: W. H. Freeman and Co., 1978), p. 2.

28. Department of Commerce, *Statistical Abstract of the United States, 1985* (Washington, D.C.: Bureau of the Census, 1984), p. 74; Cairns, *Cancer*, p. 10. Joseph L. Lyon et al., "Cancer Incidence in Mormons and Non-Mormons in Utah, 1966–1970," *New England Journal of Medicine*, January 15, 1976, pp. 129–133.

29. AEC, *Fourteenth Semiannual Report*, July 1953, CIC no. 32509, p. 49; AEC General Manager A. R. Luedecke to Nathan Woodruff, director of the Division of Operational Safety, January 11, 1962, memorandum

containing enclosure 1, "History of Radiological Safety Criteria for the Nevada Test Site."

30. Graves in *Bulloch*, vol. 1, p. 767; Gordon Dunning, TT, p. 4372; defendant's closing arguments, TT, p. 113.

31. John G. Fuller, *The Day We Bombed Utah: America's Most Lethal Secret* (New York: New American Library, 1984). The evidence and testimony from two trials in federal court, government documents, newspaper accounts, one congressional hearing and its extensive record, and the Fuller book cover the sheep deaths and injuries in overwhelming detail.

32. Elma Mackelprang Barnett, TT, pp. 1948–1969; Stephen L. Brower, TT, p. 236.

33. Frank Butrico, TT, pp. 836–837.

34. Rudgar C. Atkin, former temple and stake president (positions of importance in the local Mormon church hierarchy), interview with author, St. George, May 2, 1983; Hazel Bradshaw, ed., *Under Dixie Sun: A History of Washington County by Those Who Loved Their Forebears* (St. George: Washington County Chapter, Daughters of Utah Pioneers), p. 24.

35. Department of Energy, "Survey of Life-styles, Food Habits, and Agricultural Practices," Nevada Operations Office, 1981. Actually, a larger number than indicated drank raw milk, since many bought such milk from others.

36. *U.S. News & World Report*, June 28, 1957, pp. 79–82; Arthur F. Bruhn, *Your Guide to Southern Utah's Land of Color* (Salt Lake City: Wheelwright Lithographing Co., 1952), p. 20.

37. *Washington County News*, February 5, April 2, May 7, 1953.

38. Zion National Park, Superintendent's Report, May 1953. On May 19 Superintendent Paul R. Franks was in Capital Reef National Monument, which was also under his jurisdiction, inspecting roads that were being bulldozed across the monument to AEC uranium prospecting and mining operations.

39. *Washington County News*, April 30, May 7, 1953.

40. *Washington County News*, May 21, 1953.

41. Butrico, TT, p. 839. The offsite monitor at Mesquite later commented, "It was physically impossible to carry out the assigned duties of offsite monitoring and run a roadblock operation at the same time." The monitor added at the end of his report: "The public needs more information" (handwritten notes on Georgia Department of Public Health stationery, entitled "Summary of Roadblock Operations at Mesquite, Nevada"; communications log, May 19, 1953, entry for 0730).

42. Communications log, p. 8; Butrico, TT, p. 840.

43. Department of Energy, "Discussions with Frank Butrico," Nevada Operations Office, August 14, 1980, p. 2.

44. Butrico, TT, pp. 842–846.

45. Communications log, May 19, 1953. The log entry for 0930 notes that Graves issued the order. Gordon Dunning, who was at the test site, insists that it was Jack Clark, Grave's deputy (personal communication). I have stuck with the written record and assumed that if Clark did issue it, the order originated with his boss.

46. Butrico, TT, pp. 846, 847, 849; DOE, "Discussions."

47. DOE, "Workshop," vol. 1, June 1980, p. 144.

48. E. Elbridge Morrill, Jr., TT, pp. 2677–2684. In his subsequent report on the uranium mines, Morrill said he found "high concentrations of radioactive dust and gas" in the mines and recommended better ventilation systems. In subsequent years some uranium miners would file suit, claiming that their cancers had been caused by such conditions (E. Elbridge Morrill, "Radioactive Dust and Gas in the Uranium Mines of Utah," *American Industrial Hygiene Association Quarterly*, December 1954, pp. 271–276).

49. JoAnn Taylor Workman, DP, December 4, 1981, pp. 13–14; JoAnn Workman, TT, pp. 43–45; JoAnn Workman, affidavit, May 14, 1981; Butrico, TT, p. 854; Butrico, "Discussions," pp. 6, 11; Frank Butrico, DP, October 29, 1981, pp. 3–4; Glasstone and Dolan, *Effects of Nuclear Weapons*, p. 418; defendant's answer to plaintiffs' interrogatory, *Bulloch*, JH, vol. 1, p. 720; NOAA, "Analysis," April 1981, pp. 12–13.

50. Workman, TT, pp. 42, 45–58; Workman, affidavit; Workman, DP, p. 14.

51. Kent Anderson, TT, pp. 79–84.

52. Butrico, TT, pp. 851–869; Butrico, "Discussions," pp. 2, 3, 6.

53. *Kane County Standard*, March 13, April 10, 24, May 8, 15, 22, 1953; Adonis Findlay Robinson, ed., *History of Kane County* (Kanab, Utah: Kane County Daughters of Utah Pioneers, 1970), p. 263.

54. Leah Jackson, TT, pp. 1972–1987.

55. Butrico, TT, p. 852; communications log, May 19, 1953, entry for 1425.

56. AEC/DOD, Test Information Office, press release no. 71, May 19, 1953.

57. Butrico, TT, pp. 853–854; communications log, May 19, 1953, entry for 1115; Frank Butrico to William Johnson, "Report on Sequence of Events in St. George, Utah, Following Shot Harry," memorandum, May 30, 1953. Butrico later claimed that parts of the report were falsified and stated that it was impossible for him to be at the mining camp. I am assuming that he was not there and that others from the AEC or the Public Health Service (PHS) were and that this was their report. The communication logs seem to confirm this assumption (Pope A. Lawrence, Senior Public Health Service Officer, to William S. Johnson, "Report of Telephone Conversation with

City Manager of Cedar City, Utah, and Schedule of Events that Followed,"
memorandum, May 19, 1953).

58. Butrico, TT, pp. 870–874; DOE, "Discussions," August 1980.

59. Ibid., pp. 3, 13, 18.

60. *Washington County News*, May 21, 1953; *Iron County Record*, May
21, 1953; *Kane County Standard*, June 6, 1953.

61. *Las Vegas Review-Journal*, May 21, 1953; *Deseret News*, May 21,
1953; *Salt Lake Tribune*, May 21, 1953.

62. *Los Angeles Times*, May 20, 1953; *San Francisco Examiner*, May 20,
1953; *New York Times*, May 20 and 25, 1953.

63. Gladwin Hill, letter to author, December 26, 1982. Hill added, "As I
say, in today's crossfire of argumentation about every aspect of atomic energy, the setting and dynamics back then seem a little weird, and indeed
they were—a sort of informational semivacuum preceding the hurricane."

64. Douglas R. Stringfellow to Gordon E. Dean, Chairman of the AEC,
May 20, 1953, CIC no. 18851; Gordon Dean diary, May 25, 1953, JH, vol.
1, p. 149; Stringfellow to Dean, June 8, 1953, AEC.

65. Arthur V. Watkins to Gordon E. Dean, May 23, 1953, AEC; Dean
to Watkins, June 8, 1953, AEC.

66. Gordon Dunning to unnamed citizen, June 12, 1953, AEC.

67. AEC, minutes of meeting, May 21, 1953, JH, vol. 1, pp. 143–144;
AEC, minutes of meeting, May 22, 1953, p. 146; AEC, minutes of meeting,
May 26, 1953, p. 152; minutes of Eighty-first Conference, AEC-Military
Liaison Committee, May 28, 1953; AEC, minutes of meeting, June 17, 1953;
AEC, "Fourteenth Semiannual Report," July 1953, pp. 53–54.

68. Communications log, entry for May 20, 1953; Butrico, TT, pp.881,
903–905; Butrico memorandum, May 30, 1953; William S. Johnson, TT, pp.
4607–4608, 4643–4644; DOE, "Workshop," vol. 2, June 1980, p. 84.

69. Butrico, TT, pp. 883–884; DOE, "Discussions," p. 3.

70. DOE, "Proceedings," p. 169; DOE, "Discussions," p. 20.

71. Author's visit to test site, April 12, 1983.

72. Energy Research and Development Administration, "Final Impact
Statement, Nevada Test Site, Nye County, Nevada," September 1977, pp.
2–13; Melvin W. Carter and A. Alan Moghissi, "Three Decades of Nuclear
Testing," *Health Physics*, July 1977, pp. 55–71; Jack Dennis, ed., *The
Nuclear Almanac: Confronting the Atom in War and Peace* (Reading, Mass.:
Addison-Wesley Publishing Co., 1984), pp. 304–305, 310–311. The United
States detonated more nuclear explosions during the years of atmospheric
testing than the Soviet Union. However, the total yield of the Soviet tests
was greater and the Russians moved their tests around more. The British,
French, Chinese, and Indian tests did not come close to matching the activity of the two superpowers. Since 1962 the British have conducted their
tests at the Nevada Test Site.

73. Testimony of Bruce W. Church, TT, Director, Health Physics Division, Nevada Operations Office, pp. 4847, 4850; *U.S. News & World Report*, November 20, 1961, pp. 48–51; DOE, "Announced Tests," January 1983.

74. Statement of Philip W. Krey, Director, Analytical Chemistry Division, DOE Environmental Measurements Laboratory, Dose Assessment Advisory Group meeting, Las Vegas, January 7, 1983.

Chapter 1: The Discovery

1. DOE, "Workshop," vol. 3, p. 9; Thomas H. Saffer and Orville E. Kelly, *Countdown Zero* (New York: G. P. Putnam's Sons, 1982), pp. 149–156; A. Costandina Titus, *Bombs in the Backyard: Atomic Testing and Atomic Politics* (Reno: University of Nevada Press, 1986), pp. 107–108. The Board of Veterans Appeals finally reversed its third denial of Cooper's claim but never admitted that his disease may have been caused by the test. Cooper died in February 1978, but his outcry echoed after him.

2. *Deseret News*, August 12, 1977, February 13, 1979.

3. Vonda McKinney, TT, p. 2189; *Phoenix Gazette*, December 16, 1977; *Arizona Daily Star*, August 17, 1980.

4. McKinney, TT, p. 2200.

5. Vonda McKinney, interview with author, Holbrook, Ariz., April 27, 1983; Vonda McKinney, DP, September 24, 1980, pp. 49, 50.

6. Dale Haralson interview, Salt Lake City, September 14, 1982, and Tucson, Ariz., January 19, 20, 21, 1983.

7. Haralson interview; Haralson résumé, undated; *Tucson Citizen*, May 5, 1982; Burton J. Kinerk (Haralson's law partner), interview with author, January 20, 1983; *Journal of the Association of Trial Lawyers of America*, vol. 36, 1976, p. 266.

8. The attorney customarily gets one-third the award under state law. Under the Tort Claims Act, a federal law, the lawyer or lawyers for the plaintiffs get one-fourth the award if the case goes to trial and is successful. In both cases the cut is higher if the case is appealed or if there is a new trial. If the suit is lost, under such a contingency fee arrangement the lawyer gets nothing. It is a gamble, and the lawyer has to weigh the odds carefully. Haralson did just that. (Barber, Haralson & Kinerk, "Agreement With Attorney for Contingent Fee," undated form used by Haralson for the Allen case). There were repeated appeals from the lawyers to the clients for twenty-five dollars to help defray the expenses.

9. Haralson interview; McKinney interview; notes made by Haralson of meeting.

10. Haralson and McKinney interviews; Rose Mackelprang and VeRene Tait interviews, Fredonia, Ariz., April 24, 1983; VeRene Tait, DP, September 24, 1980; McKinney, DP, p. 53; McKinney, TT, p. 2194.

11. Haralson interview.

12. In a December 18, 1956, memorandum from Morse Salisbury, director of the AEC's Division of Information Services, to AEC commissioner Willard F. Libby, Salisbury discussed the possibility of a Disney movie on fallout. Salisbury wrote, "Unquestionably, Disney has a strong interest in atomic energy. However, he has never shown a tendency to tackle controversial subject matter. For example, his new film and book on atomic energy, 'Our Friend the Atom,' do not mention fallout." The AEC did not pursue the idea because it "smacks a little too much of propaganda," Libby wrote in a letter of December 20, 1956, to Nicholas Metropolia at the Los Alamos Scientific Laboratory, who had originally suggested the movie idea.

13. Haralson interview.

14. Haralson note for leukemia files, October 12, 1978.

15. Gofman, TT, pp. 3390–3400.

16. John Gofman, Professor of Medical Physics, Donner Laboratory, University of California, speech to Public Relations Society of America, June 2, 1957; Gofman, TT, pp. 3504, 3621; Atomic Energy Commission, "Staff Report on Allegations Made by Drs. Tamplin and Gofman of Censorship and Reprisal by the Atomic Energy Commission and the Lawrence Radiation Laboratory at Livermore," July 21, 1970; Gofman, TT, p. 3621.

17. Gofman, *Radiation and Human Health*, p. 55. For Gofman's explanation I relied not only on his court testimony and book but also on various other standard reference works, including those published by the AEC, DOD, and National Academy of Science.

18. Haralson notes. The case of Charles W. Mays, Research Professor of Pharmacology and Adjunct Professor of Physics at the University of Utah, is a curious one. When contacted by Haralson, he offered to work for the plaintiffs for forty dollars an hour and actually charged Haralson two hundred dollars for answering some questions. He was sent the money. I saw him give government attorneys at the Allen trial material that helped refute the testimony of one of the plaintiffs' witnesses. In an interview on May 10, 1983, I asked Mays about his dual role. He confirmed that he had received the two hundred dollars and had given the government lawyers an article on beagle experiments. He said that he had "tried to stay neutral in the Irene Allen case." Gofman and Morgan, who were retired at the time of the trial, charged the plaintiffs for their services, as did others.

19. Tait interview.

20. McKinney, Tait, and Mackelprang interviews; Haralson interview; typed notes, "Summary of Decedents—Leukemia Claims," Haralson files, undated.

21. *Washington Post*, September 28, 1978; *Arizona Daily Star*, September 29, 1978.

22. Stewart L. Udall, interviews with the author, Phoenix, Ariz., December 1, 1981, January 17, 1983, and Salt Lake City, September 17, 1982; "Udall's Fallout Fight," *National Law Journal*, March 19, 1979, p. 1.

23. *Washington Post*, July 2, 1978. Curry followed up on this story with a number of other stories in the *Post* and, when he changed jobs, in the *Los Angeles Times*. Other writers from national publications and networks came and went, but Curry persisted longer than others at that level. As the issue heated up, A. O. Sulzberger, Jr., wrote at least two major stories for the *New York Times*. At the regional level, Joe Bauman of the *Deseret News* and Mary Manning of the *Las Vegas Sun* pursued the story for a long time. Newspapers in Boston, Chicago, Washington, Salt Lake City, Las Vegas, Reno, and Tucson ran series on the subject, and there were articles in *Time*, *Newsweek*, *Life*, *People*, and *Parade* magazines, among other publications. The subject was covered extensively by KUTV in Salt Lake City and to a lesser extent by the national television networks. "Good Morning America" did a four-part series. Ted Koppel hosted a "Nightline" report entitled "Nuclear Fallout—Did the Government Lie?" Foreign journalists trooped to St. George, where Irma Thomas, who was frequently interviewed, kept a guest book that was a who's who of international journalism. Three books were written on the soldiers, one on the sheep, and four (including this book) on the offsite civilians. The coverage peaked in 1979 when congressional hearings were conducted and massive amounts of previously classified documents were made available, again in 1982 when the trial was held in Salt Lake City, and finally in the mid-1980s when most of the books were published. Despite the massive coverage, it is interesting how quickly an issue like this can disappear from the public consciousness.

24. *Nevada State Journal* and *Reno Evening Gazette*, April 29, 1979; Udall interview.

25. Mackelprang interview; Udall interview; Haralson interview; Rose Mackelprang to Mrs. Paxton, legal aide to Haralson, August 26, 1978.

26. Udall interview.

27. Udall interview; Congressmen Dan Marriott and Gunn McKay of Utah and Senators Orrin Hatch and Jake Garn of Utah and Barry Goldwater and Dennis DeConcini of Arizona to President Jimmy Carter, October 5, 1978.

28. *Deseret News*, September 29, October 3, 4, 1978.

29. J. MacArthur Wright, interviews with author, St. George, December 1981 and August 1982.

30. *Deseret News*, September 26, October 5, 30, December 21, 23, 1978.

31. Haralson interview; Udall interview; Udall-Haralson memo, "Initial Division of Labor," October 19, 1978; Udall to Haralson, October 19,

1978; Haralson to his law partners, memorandum, October 23, 1978; Udall-Haralson, "Position Paper—Leukemia Cases," undated.

32. Haralson to Udall, May 5, 1979.

33. *Deseret News*, October 30, December 21, 1978; Wright interview.

34. *National Law Journal*, March 19, 1979, p. 1; Wright interview.

35. *Spectrum*, January 17, 18, 19, 1979; *Deseret News*, January 15, 1979; field comments taken April 20, 1979, at public hearing held by Senator Jake Garn in St. George, JH, vol. 1, pp. 438–441.

36. *Deseret News*, December 20, 1980; *Deseret News*, January 3, 1981; Phillip S. Johnson, Assistant Professor of Sociology, Dixie College, interview with author, St. George, Utah, January 1983.

37. Glyn G. Caldwell, TT, pp. 2085, 2096, 2097.

38. Haralson notes on telephone conversations with Caldwell on August 29, 1978, and the following month; Caldwell to Haralson, September 1, 1978, with enclosures.

39. Haralson notes of telephone call to Knapp, October 12, 1978; Knapp to Haralson and Udall, January 20, 1982; Haralson notes on undated telephone conversation with Knapp.

40. Haralson notes of meeting with Knapp, October 19, 1978; Haralson notes on telephone call with Knapp, October 12, 1978; Haralson interview.

41. Harold Knapp to L. Joe Deal, Assistant Director for Field Operations, Division of Operational and Environmental Safety, DOE, January 5, 1979.

42. Haralson's notes of telephone conversation with Mays, September 1, 1978, and subsequent conversations; Mays to Haralson, September 2, 1978; Mays to Haralson, statement for services, October 13, 1978.

43. Haralson notes of telephone conversation with K. Z. Morgan, September 29, 1978; Haralson notes of visit with Morgan, August 11, 1981; Haralson interview; Morgan, TT, pp. 2721–2731; K. Z. Morgan, interview with author, Washington, D.C., March 16, 1983.

44. Ernest J. Sternglass, *Secret Fallout: Low-Level Radiation from Hiroshima to Three-Mile Island* (New York: McGraw-Hill Book Co., 1981), p. 195.

45. Haralson's undated notes of telephone conversation with Sternglass; Haralson memo for files on Sternglass meeting, October 17, 1978; Haralson interview; Haralson memorandum on Tamplin meetings, March 7, 1979, October 18, 1982. If there should be a big judgment, Sternglass told Haralson, perhaps a donation could be made to the University of Pittsburgh (Haralson memorandum on Tamplin meetings, March 7, 1979, October 18, 1982). The compensation of time and expenses is understandable in such a case, just as the different views toward compensation are quite interesting.

46. Haralson notes on telephone call with Lyon, August 8, 1978; Haralson to Lyon, February 22, 1979; Haralson interview.

47. Lyon to Tom Hanson, NIH, May 17, 1979.

48. Notes of Morgan conference, October 31, 1978.

49. Haralson notes of telephone conversation with Dr. Alice Stewart, January 7, 1981; Haralson interview.

50. Haralson to Pauling, July 10, 1981; Haralson interview.

Chapter 2: The Bureaucracy

1. Malon E. Gates to Duane C. Sewell, Assistant Secretary for Defense Programs, October 13, 1978.

2. Department of Energy, "Fiscal Year 1984 Budget Highlights," Office of the Controller, January 1983, p. 3.

3. Department of Energy, "Discussion Paper: Off-Site Radiation Exposures," Nevada Operations Office, November 11, 1978.

4. Leon Silverstrom, Chief Counsel for Nevada Operations Office, statement at Dose Assessment Steering Group Committee meeting, October 10, 1979; Walter H. Weyzen, Manager of Human Health Studies Programs, DOE Office of Health and Environmental Research, to W. W. Bun, director of the office, memorandum, December 19, 1978.

5. L. Joe Deal, Assistant Director for Field Operations, to Hal L. Hollister, Acting Deputy Assistant Secretary for Environment and Research, memorandum, February 14, 1979.

6. Maj. Gen. J. K. Bratton, Director of Military Applications, to Gates, March 3, 1979; Gates's letter to all field offices under his jurisdiction, in reference to Bratton's order, June 4, 1979.

7. *Bulletin of the Atomic Scientists*, March 1980, p. 22. When I visited the library a few years later, the staff was still edgy about the experience.

8. Gates to Duane C. Sewell, Assistant Secretary for Defense Programs, memorandum, October 13, 1979.

9. This was my impression after attending two meetings, reading the transcripts of all the meetings, informally talking to the members, reading the court testimony of the chairman, Robert D. Moseley, and the project director, Bruce W. Church, and interviewing Church. The members were either generalists who were uninterested, uninformed, and did not attend or specialists who were known in their various fields, unable to spend much time studying the voluminous paperwork produced by the staff, prejudiced to begin with because of their close associations with the government on nuclear matters, or conscientious and informed. When the secretary of the Department of Energy nearly abolished DAAG in 1981, the argument that

gave it a renewed lease on life was: "The work being done by the DAAG and the subtask groups of the project is being utilized by the Office of General Counsel" for developing a legal defense in the Allen case (K. Dean Helms, DOE Advisory Committee Management Officer, to Secretary, DOE, memorandum, September 18, 1981). Seymour Jablon of the National Academy of Sciences and a DAAG member said, "But let's face it. What drives this effort and what is driving other efforts are certain political problems and certain litigation and certain epidemiological studies that have been published and that are being done, and this effort has got to serve the need so generated." The impression that was given to outsiders was that DAAG ran the show, when at most the advisory group made suggestions that could be taken or ignored by the DOE staff, whose primary effort was to provide evidence of low doses. An advisory group is advisory at best and is usually formed to reinforce the position of those being advised. No advisee wants a runaway advisory group, and this project was no exception to that general rule (Robert D. Moseley, Chairman of DAAG, TT, pp. 6319–6320, 6329).

10. Joseph DiStefano, Acting Assistant General Counsel for General Litigation, DOE, to Stewart L. Udall, March 1, 1979.

11. Four days of joint hearings were held by a combination of two subcommittees and one committee of the House and Senate in the spring and summer of 1979. Two days of hearings were conducted in Washington, D.C., and one day each in Las Vegas and Salt Lake City. The subsequent record, printed in three volumes, ran to 5,400 pages. Although poorly organized, the record is second only to the Coordination and Information Center as a source of information on this subject. The documents in the Congressional archives and those housed in the CIC overlap in many instances (JH, vols. 1, 2, 3).

12. Testimony of Rex Lee, JH, vol. 3, pp. 240, 252, 253.

13. Notes of meeting taken by Jon T. Brown of Udall's office, February 13, 1979.

14. Haralson notes of meeting; Udall associate Stephen Roady's notes of meeting; Jon Brown's notes of meeting; Udall to Haralson and Wright, memorandum, March 14, 1979; William G. Schaffer, interview with author, Washington, D.C., March 7, 1983.

15. Udall et al., "Can the Judicial System Handle the Atmospheric Fallout Claims: A Preliminary Examination of the Negligence Issues Which Will Arise Under the Federal Tort Claims Act," April 10, 1979.

16. Plaintiffs' notes of meeting, possibly taken by Jon Brown; Schaffer interview.

17. Roady's notes of meeting, April 17, 1979.

18. Udall memorandum for the files, May 16, 1979.

19. "Memorandum for the President from Governor Matheson," Novem-

ber 21, 1978; Michael Zimmerman, interview with the author, Salt Lake City, May 16, 1983. Zimmerman was an aide to Matheson on the fallout issue at this time.

20. Matheson to the secretaries of HEW, DOE, DOD, and Agriculture, and the administrator of the Environmental Protection Agency, dated variously in January 1979.

21. Memorandum for the secretary of HEW and accompanying White House press release, November 27, 1978; Califano, memorandum for the Assistant Secretary of Health, January 5, 1979; Haralson interview; Haralson notes of telephone conversation with Curry, January 5, 1979; *Washington Post*, January 8, 1979; Matheson to Califano, January 9, 1979. Two of Matheson's aides, Zimmerman and Maggi Wilde, visited Washington to press for a new health study, a possibility that Carter had mentioned in his press release and in his instructions to Califano. They got nowhere at the National Institutes of Health. An anonymous woman caller reached Zimmerman in his hotel room, identified herself as being from NIH, and told him they were getting the runaround. "The people in the NIH are not supportive of the study," she said. Zimmerman, a Salt Lake City lawyer who also served on DAAG as the governor's representative and was later named to the state supreme court, thought the federal health bureaucracy was never very interested in the issue and only waited until Califano left office to forget it (Zimmerman interview).

22. "Report of the Interagency Task Force on the Health Effects of Ionizing Radiation," June 1979, EV, defendant's exhibit no. 638, pp. 18, 33, 73–74, 84, 100, 101.

23. Testimony of F. Peter Libassi, JH, vol. 1, pp. 99, 100.

24. Udall to plaintiffs' attorneys, memorandum, June 25, 1979.

25. Udall to plaintiffs' attorneys, memorandum, July 11, 1979.

26. Town meeting held by Senator Orrin Hatch in St. George, April 17, 1979, transcript, pp. 26–28.

27. Task Force, June 1979, p. 96; Stuart Eizenstat, Assistant to the President for Domestic Affairs and Policy, to the Attorney General, memorandum, July 20, 1979.

28. Schaffer interview; Susan Q. Stranahan, "Downwind Deals," *Alicia Patterson Reports*, 1984, p. 17.

29. Task Force on Compensation for Radiation-Related Illness, minutes of meeting, August 8, 1979.

30. Schaffer interview.

31. *Deseret News*, March 6, 1980; *New York Times*, March 19, 1980; *Deseret News*, December 31, 1980.

32. "Report of the Interagency Task Force on Compensation for Radiation-Related Illness," February 1, 1980. The no-threshold theory, which

replaced the threshold theory, holds that cancer can be induced by any amount of radiation. My copy of the report contains the stamped name and address of Gordon Eliot White, the *Deseret News* reporter who unearthed it. I treat it as a rare document.

33. Ibid., pp. 9, 10, 12, 31.

34. Ibid., pp. 39, 41–43, 50, 52, 56, 57.

35. Udall to plaintiffs' lawyers, March 3, 1980; Schaffer interview. Udall's legal views were frequently off the mark, but in looking forward to the inevitable appeal of the Allen suit, he correctly guessed that they would have difficulty with the Denver federal appeals court on the discretionary function clause of the Federal Tort Claims Act. He said that the Tenth Circuit Court had "the most extreme pro-government record" in the country on those types of rulings.

36. DOE, DNA, and DOD letters to Eizenstat, April 1980. Eizenstat to Kennedy, August 19, 1980. The agencies agreed with the broad conclusions but asked that some technical and scientific statements be reviewed. They noted that the task force was mostly composed of lawyers.

37. Schaffer interview.

38. Statement of Griffith before the Senate Committee on Labor and Human Resources, October 27, 1981; statement of Griffith before the Senate Agency Administration Subcommittee, the Judiciary Committee, and the Labor and Human Resources Committee, March 12, 1982.

39. Defense Nuclear Agency, Public Affairs Office, "Fact Sheet," July 1, 1981; visit to DNA public affairs office, headed by Lt. Col. Dale Keller, and interview with Robert L. Brittigan, general counsel, March 21, 1983. Brittigan repeated what his boss, Griffith, said and added that the claims had to be contested because of the concern of our NATO allies, the pressure from the antinuclear movement to withdraw nuclear weapons from Europe, and the difficulty nuclear-powered and armed U.S. warships had in docking in various countries—a jump in logic that I was not able to follow.

40. Troy E. Wade II, interview with author, Washington, D.C., June 18, 1983.

Chapter 3: Congress

1. Senate Committee on Labor and Human Resources, Radiation Exposure Compensation Act of 1982—Part 2, 97th Cong., 2d sess., 1982, p. 10.

2. Howard Ball, *Justice Downwind: America's Atomic Testing Program in the 1950s* (New York: Oxford University Press, 1986), pp. 170–175; A. Costandina Titus, *Bombs in the Backyard: Atomic Testing and American Politics* (Reno: University of Nevada Press, 1986), pp. 131–135. The United States has been inconsistent in deciding who was to be compensated. Be-

sides the Marshall Islanders, the United States compensated the family of one of the Japanese fishermen who died as a result of the fallout from Shot Bravo in 1954 and agreed to pay $2 million to the Japanese for loss of property and life as a result of that incident (Robert A. Divine, *Blowing on the Wind: The Nuclear Test Ban Debate, 1954–1960* [New York: Oxford University Press, 1978], pp. 30–31).

3. Ron Preston and Ed Darrell, staff members of Senate Committee on Labor and Human Resources, interviews with author, February 1983.

4. Statements of Senators Kennedy and Hatch, JH, vol. 1, pp. 2, 7.

5. *Washington Post*, April 20, 1979.

6. House Subcommittee on Oversight and Investigations, *The Forgotten Guinea Pigs*, 96th Cong., 2d sess., August 1980, pp. 33, 36–37.

7. Congressional Record, Senate, October 9, 1979, pp. 14251–14255.

8. Ron Preston to Hatch, "Utah Radiation Problem—Full Status Report," memorandum, June 25, 1979.

9. *Congressional Record*, Senate, July 27, 1979; *Washington Post*, June 29, 1979.

10. Alan A. Parker, Assistant Attorney General, to James T. McIntyre, Jr., Director of the Office of Management and Budget, June 2, 1980; Togo D. West, Jr., General Counsel of the Department of Defense, to Sen. Edward M. Kennedy, June 9, 1980.

11. The six senators to their colleagues, July 8, 1981; Preston interview.

12. Press release, office of Sen. Orrin Hatch, July 15, 1981.

13. Senate Committee on Labor and Human Resources, Radiation Exposure and Compensation Act of 1981, 97th Cong., 1st sess., 1982, pp. 242–257.

14. Troy E. Wade II to Sen. Orrin G. Hatch, April 16, 1982; Lt. Gen. Harry A. Griffith to Sen. Orrin G. Hatch, April 16, 1982.

15. Charles W. Mays, "How Many Thyroid Cancers Might Qualify for Compensation Under the Atomic Bomb Fallout Bill," unpublished study done for staff of Senate Committee on Labor and Human Resources, May 20, 1982; Elizabeth Wehr, "Senate Committee Reports Measure to Help Radiation Victims Win Damage Suits," *Congressional Quarterly*, April 24, 1982, p. 954; Sen. Howard M. Metzenbaum to Sen. Orrin G. Hatch, May 3, 1982; Hatch to Metzenbaum, May 14, 1982; *Salt Lake Tribune*, May 7, 1982; *Deseret News*, May 7, 1982; Elizabeth Wehr, "Bill to Help Radiation Victims Collect Damages Bogs Down," *Congressional Quarterly*, July 3, 1982, p. 1588; Preston and Darrell interviews.

16. *Salt Lake Tribune*, August 29, 1982.

17. Robert A. McConnell, Assistant Attorney General, to Sen. Orrin G. Hatch, December 13, 1982; Hatch to William French Smith, Attorney General, December 17, 1982; *Washington Post*, January 3, 1983; White House,

Office of the President, Statement of the President on the Orphan Drug Act, January 4, 1983.

18. Committee on the Biological Effects of Ionizing Radiation, *The Effects on Populations of Exposures to Low Levels of Ionizing Radiation: 1980*, p. 136.

Chapter 4: Lawyers and the Law

1. Haralson interview.

2. Henry A. Gill, Jr., interview with author, Salt Lake City, November 11, 1982, and Washington, D.C., March 7, 1983.

3. Kinerk interview.

4. Ralph Hunsaker, interview with author, Salt Lake City, November 6, 1982.

5. Haralson to Hunsaker, June 9, 1982. The final disposition of the fee was worked out in the following manner: the lawyers would collect 25 percent of the award, if there was any, and of that amount Udall would get 40 percent and Wright, Hunsaker, and Haralson 20 percent each.

6. Hunsaker and Gill interviews.

7. Jeffrey Axelrad, Director of Torts Branch, Department of Justice, interview with author, Washington, D.C., March 7, 1983; Gill interview.

8. Stewart L. Udall to J. Paul McGrath, "A Proposal to Settle Some of the Fallout Cases," memorandum, June 25, 1982; Udall interview.

9. Axelrad interview.

10. Gill and Udall interviews.

11. The following discussion of the Federal Tort Claims Act is based mainly on two books that are well known to students of the law: Lester S. Jayson, *Handling Federal Tort Claims: Administrative and Judicial Remedies* (New York: Matthew Bender, 1982); William L. Prosser, *Handbook of the Law of Torts* (St. Paul: West Publishing Co., 1971). Jayson, a former director of the Torts Section of the Department of Justice, testified at the 1979 congressional hearings on the problems of the Federal Tort Claims Act. In addition, Wayne McCormick and Fred Anderson of the University of Utah School of Law helped me with specific questions. The voluminous pleadings at the pretrial, trial, and posttrial levels added to the available material on the subject. The law does not lack for documentation.

12. Jayson, *Federal Tort Claims*, p. 54.

13. Jayson testimony, JH, vol. 3, p. 246.

14. Haralson to Jane W. Bradshaw (and others), August 29, 1980. Earlier that month Vonda McKinney had been quoted in the *Arizona Daily Star* as stating that her mother-in-law suspected radiation as the cause of her son's cancer all along. "We wondered if the beautiful mushroom-shaped

clouds could be killing our families," the reporter quoted Vonda McKinney as saying (*Arizona Daily Star*, August 17, 1980). Mrs. McKinney later said that she had been misquoted.

15. Prosser, *Law of Torts*, p. 974.
16. Jayson, *Federal Tort Claims*, pp. 54, 245.
17. Ibid., p. 146.
18. Benjamin N. Cardozo, *The Nature of the Judicial Process* (1921; New Haven: Yale University Press, 1949), p. 166.

Chapter 5: The Site

1. Senate Committee on Public Buildings and Grounds, "Report on S. 2178 to Build a Public Building in Salt Lake City and Ogden," April 20, 1896.

2. League of Women Voters of Salt Lake City, questionnaire, October 1, 1959; *Deseret News* Legislative Biographical Sheet, 1965; *Deseret News*, March 30, 1952, March 11, 1962, January 6, 14, 1965, May 23, 1971, August 29, 1978; Judge Bruce S. Jenkins, speech to Utah State Bar, Salt Lake City, January 16, 1981; Jenkins, speech at graduation exercises, Highland High School, Salt Lake City, June 4, 1982; Jenkins, speech at admission exercises to Utah State Bar, May 5, 1982; Jenkins to author, December 10, 1982. Traditionally, the nominee for a Utah federal judgeship is recommended by the state's senior senator, but since both senators were Republicans and President Carter was a Democrat, a selection committee was formed that came up with five names. Rex E. Lee, who in 1978 was dean of the law school at Brigham Young University, headed the committee, which submitted five names, all of whom were acceptable to Senators Garn and Hatch. Jenkins, who had long been active in Democratic party matters in Utah, was picked by Carter over the other four, who included Dan Bushnell, the lawyer for the sheepmen. Shortly after confirmation, Jenkins won the draw for the Allen suit (*Deseret News*, April 8, 1978).

3. Structurally, the law marches to a different drummer, so the order of the material has been altered in order to tell a coherent story in the manner described.

4. Ernest O. Lawrence, Professor of Physics, communication to unknown recipient on February 27, 1932; Norris E. Bradbury, interview with Arthur Laurence Norberg, Oral History Project for the History of Science and Technology Program, Bancroft Library, University of California, 1980; Norris E. Bradbury, interview with the author, Los Alamos, N.M., April 25, 1983; Norris E. Bradbury, TT, pp. 5968–5974.

5. Bradbury, TT, pp. 5970–5971; George B. Kistiakowsky, *Bulletin of the Atomic Scientists*, June 1980, pp. 19–20; Richard Rhodes, *The Making of the Atomic Bomb* (New York: Simon and Schuster, 1986), pp. 657–664; Bradbury, Norberg interview, p. 38; Richard G. Hewlett and Francis Duncan, *Atomic Shield* (Washington, D.C.: Atomic Energy Commission, 1972), pp. 319, 378.

6. Howard L. Andrews, "Radioactive Fallout from Bomb Clouds," *Science*, September 9, 1955, p. 455; Howard L. Andrews, "Residual Radioactivity Associated with the Testing of Nuclear Devices within the Continental Limits of the United States," National Institutes of Health, September 13, 1953; Glasstone and Dolan, *Effects of Nuclear Weapons*, p. 622; Edward Teller and Allen Brown, *The Legacy of Hiroshima* (Garden City, N.Y.: Doubleday & Co., 1962), p. 170. The glasslike fused sand that was drenched with radiation at the Trinity site was gathered by collectors, set in costume jewelry, and sold to the public. Los Alamos scientists gave mayors of the nation's forty-two largest cities paperweights made from the same substance. Paul Boyer, *By the Bomb's Early Light: American Thought and Culture at the Dawn of the Atomic Age* (New York: Pantheon Books, 1985), pp. 11, 63.

7. Bradbury, Norberg interview, p. 34.

8. Ibid., p. 41; Bradbury, TT, pp. 5972–5973; Atomic Energy Commission, *In the Matter of J. Robert Oppenheimer* (Cambridge, Mass.: MIT Press, 1971), pp. 482–483.

9. Bradbury, Norberg interview, p. 43; Bradbury, TT, p. 5972.

10. Hewlett and Duncan, *Atomic Shield*, p. 47; David Alan Rosenberg, "U.S. Nuclear Stockpile, 1945–1950," *Bulletin of the Atomic Scientists*, May 1982, pp. 25–30; Department of Defense, Modern Military History Division, *The History of the Joint Chiefs of Staff*, vol. 1: *1945–1947*, p. 294.

11. Ernest O. Lawrence, nominating statement for Alumnus of the Year, January 1952; Robert Jungk, *Brighter than a Thousand Suns: A Personal History of the Atomic Scientists* (New York: Harcourt Brace Jovanovich, 1958), p. 242.

12. AEC, *Oppenheimer*, p. 493. Norberg interviews, Oral History Project, Science and Technology Program, Bancroft Library, University of California: Robert M. Underhill, Secretary-Treasurer of the Regents, 1976; John H. Manley, Associate Director of Los Alamos Scientific Laboratory, 1980; Raemer E. Schreiber, Division Leader and Deputy Director, Los Alamos, 1980; Bradbury, Norberg interview, p. 52.

13. Bradbury, Norberg interview, p. 52.

14. Ibid., p. 65; "Bradbury's Colleagues Remember His Era," *Los Alamos Science*, Winter/Spring 1983, p. 33.

15. Tape recording of Old Timers Reunion, June 14, 1980.

16. Schreiber, Norberg interview; Bradbury, TT, p. 5978; Bradbury, Norberg interview.

17. Rear Adm. William S. Parsons to Lt. Gen. J. E. Hull, Commander, Joint Task Force-7, "Location of Proving Ground for Atomic Weapons," memorandum, May 12, 1948; Commander, Joint Task Force One, to Joint Chiefs of Staff, "Preliminary Technical Results," August 10, 1946.

18. Bradbury, TT, p. 5978.

19. Bradbury to Roger Warner, June 30, 1947; David E. Lilienthal, *The Journals of David E. Lilienthal*, vol. 2 (New York: Harper & Row, 1964), p. 213.

20. Bradbury, TT, p. 5979; AEC, "Report of Committee on Operational Future of Nevada Proving Grounds," May 11, 1953.

21. Bradbury, TT, p. 5988; Col. B. G. Holzman, staff meteorologist, to Rear Adm. William S. Parsons, "Site for Atomic Bomb Experiments," memorandum, April 21, 1948; Parsons to Lt. Gen. John E. Hull, May 12, 1948.

22. David E. Lilienthal, Chairman of the AEC, to Clark Clifford, an aide to President Truman, December 14, 1948; Parsons to Hull, May 12, 1948; Boyer, *Bomb's Early Light*, pp. 316–318.

23. Director of Military Applications, "Location of Proving Grounds for Atomic Weapons," September 15, 1948.

24. AEC, minutes of meeting, September 16, 1948; Lilienthal to Chairman, Military Liaison Committee, "Location of Proving Ground for Atomic Weapons," memorandum, September 24, 1948.

25. Captain Howard B. Hutchinson, "Project Nutmeg," Armed Forces Special Weapons Project, January 28, 1949.

26. Sumner T. Pike, Acting AEC Chairman, to Chairman, Military Liaison Committee, March 8, 1949; Col. Paul T. Preuss, an aide to Truman, to Navy Capt. James S. Russell, memorandum, February 2, 1949.

27. Bradbury, TT, p. 5981.

28. AEC, minutes of meeting, July 12, 1950; Gordon E. Dean to Robert LeBaron, Chairman of the Military Liaison Committee, memorandum, July 13, 1950.

29. "Anatomy of the Nevada Test Site," Los Alamos Scientific Laboratory, March 1965, pp. 4, 9, 16, 17, 25; "Nevada Test Site, Final Environmental Impact Statement," Energy Research and Development Administration, September 1977, pp. 2-13, 2-11, 2-12, 2-139. The original test site consisted of 680 square miles carved out of the bombing range in 1952 and enlarged by 1980 to 1,350 square miles. The Los Alamos booklet mentions that following the denuding of Frenchman and Yucca flats by the atmospheric tests each year, a species of tumbleweed named Russian thistle invaded the bare ground. East of the test site, in strange juxtaposition, are the Pahranagat

National Wildlife Refuge and the Desert National Wildlife Range.

30. Bradbury, TT, pp. 5991–5992; Bradbury to Brig. Gen. James McCormack, Jr., memorandum, July 21, 1950.

31. Col. George F. Schlatter, Division of Military Applications, memorandum for the record, July 25, 1950.

32. "Discussion of Radiological Hazards Associated with a Continental Test Site for Atomic Bombs," based on notes taken by Frederick Reines, August 1, 1950; *Deseret News*, February 3, 1979.

33. Louis H. Hempelmann et al. (other members of Los Alamos and Argonne National Laboratories), "The Acute Radiation Syndrome: A Study of Nine Cases and a Review of the Problem," printed by authority of the AEC in *Annals of Internal Medicine*, February 1952, p. 336. The whole-body dose given for Graves in this exhaustive article was 390 roentgens. Graves testified in the first sheep trial that he had received a dose of 200 roentgens, a number he repeated at the Joint Committee on Atomic Energy fallout hearings in 1957 (Graves, *Bulloch*, JH, vol. 1, p. 770). Bradbury gave this answer when questioned about the difference in numbers (Bradbury interview). One of the morals of this story is that top AEC officials deceived their own employees.

34. Jungk, *Thousand Suns*, pp. 195–196; Bradbury interview; Graves testimony in *Bulloch*, JH, vol. 1, pp. 770, 843; Lansing Lamont, *Day of Trinity* (New York: Atheneum, 1965), p. 277; Dexter Masters, *The Accident* (New York: Alfred A. Knopf, 1955). The Masters book is a fictionalized account of the Slotin accident. The incident is ripe for a better attempt.

35. Joint Committee on Atomic Energy, "The Nature of Radioactive Fallout and Its Effects on Man," 85th Cong., 1st sess., 1957, p. 102.

36. *Los Alamos Monitor*, July 29, 1965; Bradbury interview. Bradbury said that both Graves and Graves's father had died of heart attacks. It was a point that he emphasized (Bradbury interview). The local newspaper printed the information that it had been given by the AEC. It would be fascinating to know what the fate of all those involved in the accident and Graves's children and grandchildren has been.

37. "Report Covering the Selection of Proposed Emergency Proving Ground for the United States Atomic Energy Commission," Holmes & Narver, August 14, 1950; Gordon Dean diary, November 1, 1950; James S. Lay, Jr., Executive Secretary of the National Security Council, to the Secretaries of State and Defense, and the Chairman of the AEC, memorandum, November 14, 1950.

38. J. Carson Mark, "Weapons Testing," *Bulletin of the Atomic Scientists*, March 1983, pp. 45–51; Norris E. Bradbury to Brig. Gen. James McCormack, Jr., memorandum, August 22, 1950. Bradbury had his cake and ate it too. The Greenhouse test series was held in the Pacific in 1951,

and the Nevada site opened for business that same year. There have been tests of thermonuclear devices at the Nevada site, but they have been on a smaller scale than those tested in the Pacific (Bradbury, TT, pp. 5993–5994).

39. Robert J. Donovan, *Tumultuous Years: The Presidency of Harry S. Truman, 1949–1953* (New York: W. W. Norton, 1982), pp. 308–309. A clarifying statement issued later that day said, "Consideration of the use of any weapon is always implicit in the very possession of that weapon" (White House press release, November 30, 1950). The Donovan biography only mentions the press conference. Some time before August 1950, nonnuclear components of nuclear bombs were shipped overseas as part of a preparatory action Truman had approved (S. Everett Gleason, Acting Executive Secretary of the National Security Council, to President Truman, "Re: Advice for Use of Atomic Weapons," memorandum, October 23, 1952. Excerpts from the Gordon Dean diary (furnished me by the personnel of the Department of Energy archives after a specific request) posed these questions on the use of atomic bombs that needed to be answered first: "What would be the effect of its use on public opinion in the United States, in allied countries, in Asia? Should we obtain UN concurrence before using it?" Additionally, Dean stated: "It was agreed at the meeting that in the event that the Joint Chiefs of Staff should suddenly make a recommendation that an atomic bomb be used at a particular place that the NSC subcommittee [consisting of the secretary of defense, secretary of state, and chairman of AEC] would meet and address itself to an agreed-upon set of questions such as those in the Arneson [State Department] memorandum. It would be only after an analysis of the situation and the answering of the questions that the Special Committee would then make a recommendation to the President. This is a significant step because it establishes a procedure whereby AEC will participate in any such decision" (Dean diary, November 10, 1950).

40. Los Alamos Scientific Laboratory, "Desirability of an Area in the Las Vegas Bombing Range to be Used as a Continental Proving Ground for Atomic Weapons," LASL Report (Lab-J-1609), November 22, 1950. The report was Appendix F in a packet of information given to the AEC commissioners.

41. James S. Lay, Jr., to Secretaries of State and Defense and the Chairman of the AEC, "Additional Test Site," memorandum, December 19, 1950; Philip J. Farley, Acting Secretary of the AEC, "Location of Proving Ground for Atomic Weapons," note by the acting secretary to the commissioners, December 20, 1950.

42. Harry S. Truman, *Memoirs of Harry S. Truman: Years of Trial and Hope* (Garden City, N.Y.: Doubleday & Co., 1956), p. 312.

43. Harry S. Truman, private diary, entry for December 9, 1950.

44. The following information from Gordon Dean's diary was furnished

to me by Department of Energy archivists when it was broadly hinted during my visit to Germantown, Md., on March 8 and 9, 1983, that if I requested the documentation for the footnotes in the *Atomic Shield*, the official AEC history, I would discover some interesting facts that had not yet come to light. It was not practical for me to request all the documentation for the many footnotes in that lengthy book, so I made a guess at which were most promising for the needs of this project and came up with this disclosure. Other than a general statement by Truman, based on an entry in his private diary, that he twice considered the "threat" of nuclear war against China and Russia (which was published in the *New York Times* of August 3, 1980), I believe this is the first detailed discussion of this grave incident. President Eisenhower ruled out the use of atomic weapons in Vietnam in 1954, President Kennedy put U.S. nuclear forces on alert during the Cuban missile crisis, President Johnson again considered the use of nuclear weapons in Vietnam, as did President Nixon, and some consideration was given to their use by the Carter administration should Russia invade Iran or Saudi Arabia in the late 1970s (*New York Times*, September 2, 1986). Should these bare incidents be the sum total of our knowledge about the imminence of nuclear warfare, then the Truman episode comes closest to being the most detailed, first-person account of what almost occurred, not on the territory of some third country, but on the Soviet and Chinese mainland. I discuss this incident in order to give a sense of how close to the brink we have come without knowing about it and to give as vivid an account as possible of the times during which the site was selected and the first series of tests was conducted in Nevada. I was unable to find any corresponding account in the Truman archives. Archivists at the Harry S. Truman Library in Independence, Mo., stated that there was no complete record of one-on-one meetings with the president. Other than what Truman said afterwards, there is no other record of these momentous events, so we have to take Dean's word for them. Dean was not a frivolous person (Hewlett and Duncan, *Atomic Shield*, p. 356).

45. Dean diary, covering events of March 26, 1951.

46. Ibid., April 5, 1951; Hewlett and Duncan, *Atomic Shield*, pp. 538–539. Dean attempted to curb the influence of the military throughout his tenure on the AEC. He later wrote: "Too frequently, in the midst of all this, we find the military suggesting not simply *what* the Commission should do but also *how* it should be done. This the Commission has resisted." One way to lessen the influence of the military was to erect the NSC subcommittee as a buffer (Gordon Dean, *Report on the Atom* [New York: Alfred A. Knopf, 1957], pp. 137, 141).

47. Dean diary, April 6, 9, 1951.

48. Ibid., April 9, 1951.

49. Diary entry covering August 31, 1951. Actually, it was not until May 25, 1953, that such an artillery weapon was tested at the Nevada site.

50. Truman diary, January 27, 1952.

51. Interviews with Truman for memoirs, undated fragment.

52. Maj. William R. Sturges, Jr., to Col. Schlatter, "Public Relations Conference Concerning Mercury," memorandum, December 20, 1950.

53. Richard G. Elliott, Director, Information Division, to Norris E. Bradbury, Director, Los Alamos Scientific Laboratory; AEC, minutes of meeting, January 2, 1951; Dean diary, January 8, 1951; Carroll L. Tyler, manager, AEC Santa Fe office, to Gen. James McCormack, January 3, 1951.

54. Dean diary, January 9–12, 1951; Rodney L. Southwick, Assistant Chief, Public Information Service, AEC, to Joseph Short, White House Press Secretary, January 1, 1951.

55. AEC, minutes of meeting, January 10, 1951; AEC, press release no. 335, January 11, 1951; *Las Vegas Review-Journal*, January 15, 1951.

Chapter 6: The Tests

1. Sheahan in *Bulloch*, JH, vol. 1, p. 1116; Dean diary, January 27, 1951.

2. Bradbury to Lawrence, January 20, 1951.

3. Dean diary, January 29, February 6, 1951.

4. DOE, "Workshop," vol. 3, pp. 4–5. Larson and his crew were collecting soil samples because the Metropolitan Water District of Southern California was worried about the radiation levels in Lake Mead, from which most of southern California's water supply came. The soil samples would establish background levels against which later radiation levels could be measured. Larson and his UCLA Atomic Energy Project also worked directly for the AEC.

5. Gaelen Felt to Alvin C. Graves, "Jangle Fallout Problems," J Division, June 28, 1951; Harry F. Schulte to Dr. T. L. Shipman, "Preliminary Report on Buster-Jangle Fallout Program," Los Alamos Scientific Laboratory, December 15, 1951.

6. Gordon Dean, memorandum for the files, January 16, 1952.

7. Text of Gordon Dean's testimony, House Appropriations Committee hearing, "FY 1953 Estimates," February 5, 1952.

8. Press briefing, Las Vegas Municipal Auditorium, April 21, 1952.

9. Lt. Col. James B. Hartgering, Rad-Safe Advisor, to Dr. Shields Warren, Director, AEC Division of Biology and Medicine, May 28, 1952; Maj. H. M. Lulejian, Off-Site Operation Officer, to Hartgering, "Radioactive Contamination at Lincoln Mine as a Result of Snapper-Easy Shot," May 15, 1952; V. B. Lamoureux to Maj. Lulejian, "Lincoln Mine and Groom Mine Post-Shot Survey," May 8, 1952.

10. Sheahan, JH, vol. 1, pp. 1144–1147.

11. Dean diary, May 16, 1952; Hartgering to Spendlove, May 29, 1952. This story is full of odd occurrences. Roy Gibbons of the *Chicago Tribune*

called Dean at about this time and, stating that he was speaking for the newspaper, suggested that it would be a good idea to drop an atomic bomb on frozen Lake Michigan the following winter. In this way midwesterners would be made aware of AEC activities. Dean demurred. He thought about the fallout cloud that would pass over the cities. However, Dean, himself a master at public relations, told Gibbons: "But why don't you drop me a line about it and I will have our boys think about it." Needless to say, no such bomb was ever dropped on Lake Michigan.

12. R. E. Thompsett et al. to C. L. Tyler, AEC, Santa Fe, "Inspection of Cattle Belonging to Floyd Lamb, Alamo, Nevada," August 18, 1952. The report, signed by the three Los Alamos investigators, stated, "The distribution and appearance of the lesions leaves little doubt that these are superficial radiation burns from fallout material." Thompsett stated in another report that Lamb had a valid complaint (Thompsett to Frank C. DiLuzzio, field manager of Los Alamos field office, memorandum, November 18, 1952; Dunning to John C. Bugher, Director, Division of Biology and Medicine, "Radioactive Fallout on Cows Near the Nevada Test Site," August 21, 1952). Dunning's report recapped the incident but left out mention of the Thompsett et al. conclusions, stating only that the cattle had "some abnormality." Dunning was establishing a pattern of filtering out the bad news when reporting to his superiors. The last document I obtained on this matter is dated July 1956, and the AEC had still not paid anything to Lamb, who had gotten tangled up in bureaucratic procedural matters. By that time the AEC was involved in the sheep deaths, and it was unlikely that any compensation would be forthcoming (Joe B. Sanders, chief, Las Vegas Branch, Test Division, to Chalmers C. King, assistant AEC General Counsel, July 9, 1956).

13. Committee on Operational Future, Nevada Proving Grounds, summary of minutes, May 11, 1953.

14. Herbert F. York, The Advisors: Oppenheimer, Teller and the Superbomb (San Francisco: W. H. Freeman & Co., 1976), pp. 133–136; Teller, Legacy of Hiroshima, pp. 60–66; Schreiber, Norberg interview, p. 35; Samuel H. Day, "The Nuclear Weapons Labs," Bulletin of the Atomic Scientists, April 1977, p. 21. The yield for the two duds was .2 and .3 kilotons, versus 11 kilotons for the next lowest and 61 kilotons for the highest-yield shots in the series (James E. Reeves, Director, Office of Test Operations, Santa Fe, "Summary of Nuclear Detonations at the Nevada Proving Grounds," December 22, 1953).

15. Bradbury to Gen. Kenneth E. Fields, Director, Division of Military Applications, memorandum, January 12, 1953; Fields to AEC, "Proposed Program for Operation Upshot," February 2, 1953.

16. Nevada Highways and Parks, June–December 1953.

17. *The Quill*, July 1953; *Salt Lake Tribune*, March 25, 1953; *Las Vegas Sun*, April 8, 1953; *Los Angeles Examiner* photograph, spring 1953 (specific date not mentioned in CIC clipping file); Titus, *Bombs in the Backyard*, p. 93.

18. Merrill Eisenbud, Director, Health and Safety Division, Los Alamos Scientific Laboratory, to Richard G. Elliott, Santa Fe Operations Office, August 24, 1953; Richard L. Miller, *Under the Cloud: The Decades of Nuclear Testing* (New York: Free Press, 1986), pp. 58, 90–91, 157–158.

19. AEC, "Public Information Program for Upshot-Knothole," note by the secretary to the commission, February 6, 1953; AEC, *Continental Weapons Tests . . . Public Safety*, March 1953, CIC no. 1413, pp. 9, 14.

20. Sheahan, JH, vol. 1, pp. 1165–1166.

21. Winona Shah, TT, pp. 87–103; *Washington County News*, March 26, 1953; Lt. Col. Tom D. Collison, "Operation Upshot-Knothole: Radiological Safety Operation," Report to the Test Director, Armed Forces Special Weapons Project (AFSWP), June 1953, EV, defendant's exhibit no. 194, pp. 42–64; "Report of Public Health Service Activities in the Off-Site Monitoring Program," Nevada Proving Ground, spring 1953, EV, defendant's exhibit no. 195, p. 22.

22. AEC, "Background Information on Continental Nuclear Tests," Test Information Office, Las Vegas, spring 1953, pp. 10, 13.

23. AEC, Office of Test Information, press release nos. 40, 41, 42, April 25, 1953. The statements by the members of Congress, Mayor Sam Yorty of Los Angeles, and a lengthy exhortation by James H. Doolittle (who led the famous World War II bombing raid on Tokyo) for the U.S. to remain ahead of the USSR in nuclear weapons were far longer than the meager information given out that day on what was going on offsite (AEC, Office of Test Information, press release nos. 45, 46, 47, April 25, 1953; AFSWP, June 1953, pp. 105–114; PHS, 1953, p. 42; *New York Times*, August 12, 1979; *Washington County News*, April 30, 1953.

24. Mohawk Association of Scientists and Engineers, newsletter, June 10, 1953; John C. Bugher to Kenneth E. Fields, "Rainout in the Troy, New York, Area," memorandum, May 14, 1953, CIC no. 18852; AEC, "Radioactive Debris from Operations Upshot and Knothole," Health and Safety Laboratory, New York Operations Office, June 25, 1954, p. 60. The report noted that small amounts of fallout from the 1953 test series were measured in Canada, Bermuda, Germany, and at Hiroshima and Nagasaki (Ibid., p. 63).

25. AEC, minutes of meeting, May 13, 1953.

26. Ralph E. Lapp, "Nevada Test Fallout and Radioiodine in Milk," *Science*, September 1962, pp. 756–758; *Bulletin of the Atomic Scientists*, February 1955; AEC, Bureau of Investigation report by Anson M. Bartlett,

262 / NOTES TO PAGES 110–114

March 21, 1955, CIC no. 32309; Borst to Dean, May 16, 1953; AEC, minutes of meeting, May 22, 1953; Bugher to Borst, July 1, 1953; Ball, *Justice Downwind*, pp. 67–69.

27. Tyler to Fields, message, May 11, 1953.

28. Dean to Lewis L. Strauss, Special Assistant to President Eisenhower for Nuclear Matters, memorandum, May 26, 1953; Richard Elliott, TT, pp. 6084–6085.

29. AEC, minutes of meetings, May 18 and 22, 1953.

30. Ibid.

31. Dean to Strauss, May 26, 1953; Dean diary, May 26, 27, 1953.

32. Col. Marcus F. Cooper to Salisbury, "Statement That Might Be Issued if Authoritative Sources Criticize Nevada Proving Grounds Operations," memorandum, August 3, 1953. The colonel said that it is "our understanding of the present policy that no reference of any nature will be made to efforts in the thermonuclear field." In such a manner does policy get handed down. The colonel also praised the photographic industry, which had incurred costs totaling $267,000 to reduce radiation effects on light-sensitive materials but had not filed claims for damages. He suggested that the industry be praised in the redrafted press release, and it was.

33. Strauss to Eisenhower, June 4, 1953; Stephen E. Ambrose, *Eisenhower*, vol. 2: *The President* (New York: Simon and Schuster, 1984), pp. 132–133; Devine, *Blowing on the Wind*, pp. 9–11.

34. Ambrose, *Eisenhower*, vol. 2, pp. 93, 171, 225, 343, 345, 400, 479, 590.

35. Cole to A. H. Gallagher of Hot Springs, Ark., June 23, 1953. At about the same time, Cole wrote to Dean: "We have been assured by your Commission that every precaution has been taken to guarantee that no harm to people, animals, or crops or water supplies will result from the continental tests" (Cole to Dean, June 6, 1953). This is called covering your tracks.

36. AEC, minutes of meeting, June 17, 1953; Salisbury to C. Herschel Schooley, Director, Office of Public Affairs, Department of Defense, memorandum, March 10, 1955.

37. Transcript of film *Atomic Test in Nevada*, TT, pp. 38–51.

38. DOE, "Workshop," June 1980, pp. 51, 126; Haralson interview. A DOE official, Bruce Church, let slip to Haralson that the meeting had been held, and the lawyer requested the transcript as part of the discovery proceedings. It is a rich source of unguarded comments.

39. Plaintiffs' opening statement, TT, pp. 36–37; undated, untitled AEC Test Information Office document outlining public relations activities.

40. Tyler to Fields, June 16, 1953; Elliott, TT, p. 180; Elliott, "The Public Relations of Continental Tests," September 23, 1953, defendant's exhibit no. 1I, p. 16. AEC internal documents never referred to Borst by

name, but instead to the "ex-AEC scientist" (Report of Committee to Study Nevada Proving Grounds, February 1, 1954, defendant's exhibit no. 1, p. 49; Donald J. Leehey, manager, Santa Fe Operations Office, to Fields, "Public Relations in NPG Region," October 28, 1954).

41. Elliott, TT, pp. 149–152.

42. Hill letter to author; Elliott, TT, p. 6065.

43. AEC, *Fourteenth Semiannual Report of the Atomic Energy Commission*, July 1953, p. 50; AEC, minutes of meeting, May 21, 1953; PHS, 1953, p. 51. The estimated average dose was 5.2 r (AFSWP, 1953, p. 126). An average dose of 4.2 r was estimated.

44. Tyler to Fields, "Preliminary Report on Continental Tests and Future Utilization of Nevada Proving Grounds," September 29, 1953; Elliott, committee secretary, to committee membership, "Summary of Actions Taken," August 10, 1953.

45. Ambrose, *Eisenhower*, vol. 2, pp. 132–133; Ralph E. Lapp, "Atomic Candor," *Bulletin of the Atomic Scientists*, October 1954, pp. 312–314, 336. Lapp wrote, "Surveying the past nine years, one finds little official candor commensurate with the magnitude of the rising hazards from modern weapons."

46. Howard L. Andrews, "Residual Radioactivity Associated With the Testing of Nuclear Devices Within the Continental Limits of the United States," September 13, 1953, pp. 7–8. Andrews, who was exposed to fallout at the 1946 Pacific tests, wrote a popular textbook with Lapp entitled *Nuclear Radiation Physics*. A biophysicist and Public Health Service officer, Andrews was the liaison between the surgeon general and the test site management and was also a member of the test manager's advisory panel at the Nevada Test Site until 1958 (Andrews testimony, TT, pp. 4482, 4485–4486, 4489, 4493).

47. Tyler to Fields, "Nevada Proving Grounds," memorandum, December 21, 1953.

48. Committee report, February 1, 1954, pp. 7, 19, 44, 53.

49. AEC, minutes of meeting, February 17, 1954, JH, vol. 1, p. 166.

50. *Deseret News*, January 19, 1955; Elliott, TT, pp. 6089–6090. Like Graves, Elliott believed the test director was exposed to 200 roentgens of radiation and advertised his exposure as such.

51. AEC, "Atomic Tests in Nevada," 1955; K. F. Hertford, manager, AEC Albuquerque office, to Brig. Gen. Alfred D. Starbird, Director of Military Applications, memorandum, January 3, 1957. The 1957 booklet printed the same assurances: "Simply stated, all such findings have confirmed that Nevada test fallout has not caused illness or injured the health of anyone living near the test site" (AEC, "Atomic Tests in Nevada," 1957, p. 15).

52. AEC press release and Strauss statement, February 15, 1955; press

releases on speeches by Commissioner Willard F. Libby released by the AEC press office, September 29, 1955, and January 19, 1956.

53. Elliott to Gen. Starbird, Director of Military Applications, "Mass Use of Animals in Nevada Tests," memorandum, November 6, 1958, EV, plaintiffs' exhibit no. 212.

54. "Transcript of Remarks by Dr. John C. Bugher," Director, Division of Biology and Medicine, at preseries press conference, Las Vegas, February 13, 1955; Office of Test Information, press release, February 19, 1955; "Continental Atomic Tests, Background Information for Observers," Nevada Test Site, spring 1955.

55. Anderson to Strauss, letter, February 21, 1955; AEC, minutes of meeting, February 23, 1955. The question of what form the minutes should take had been debated before. In 1953 Commissioner Zuckert complained to Chairman Dean that the problem with paraphrased minutes was the inability to agree on what had been said at the meeting. Zuckert said he foresaw that verbatim minutes "might limit discussion." Then he added, "Too often we have been in the position of softening the minutes so that they may reflect what the Chairman of the MLC [Military Liaison Committee] wishes he had said, rather than what was actually said at the meeting" (Zuckert to Dean, May 8, 1953). Again, it was a situation of the military men dominating the civilians. My editing was aimed at making the conversation coherent while retaining the accuracy of what was said.

56. Minutes of Ninety-sixth AEC-MLC Conference, February 24, 1955; memo for the files from Corbin Allardice, Executive Secretary of the Joint Committee on Atomic Energy, February 24, 1955.

57. Hewlett and Duncan, *Atomic Shield*, p. 131; Nichols, TT, pp. 6469, 6451–6453, 6474, 6477, 6479.

58. Fields, TT, pp. 6469, 6499, 6515.

59. Bradbury to T. H. Johnson, Director, AEC Division of Research, November 21, 1955.

60. *Newsweek*, June 25, 1956; *Deseret News*, June 12, 1956; Gordon Dunning, "Criteria for Exposure to Populations Around the Nevada Test Site Resulting from Radioactive Fallout," undated, CIC no. 15633; Dr. Thomas L. Shipman, Los Alamos Health Division Leader, to Gordon Dunning, memorandum, August 14, 1956. Dr. Shipman commented favorably on Dunning's suggested standard and then added, "Let us not forget, however, that on a number of occasions in the past we have been extremely fortunate and have been saved from a potentially serious situation by a scant margin. I am not one who believes that this kind of luck will hold good indefinitely. Neither do I feel that we have yet reached the point where we can have a firm confidence in fallout forecasting" (Joe B. Sanders, Chief, Las Vegas Branch, to Roscoe H. Goeke, Radiological Safety Advisor, Test Divi-

sion, "Draft Criteria for Exposure to Populations Around the Nevada Test Site Resulting from Radiological Fallout by Dr. Dunning," August 16, 1956, CIC no. 15632). Sanders cited the "inbreeding" among the offsite population as being greater than at other places and as a factor that Dunning had not considered (Graves to Dunning, September 7, 1956). Graves was worried about his inability to conduct tests under the new guideline ("Radiological Safety Criteria for the Nevada Site," report to the general manager by the director of Biology and Medicine, circulated to commissioners on November 12, 1956; AEC, minutes of meeting, November 14, 1956; Kenneth F. Hertford, manager, Albuquerque Operations Office, to General Starbird, Director of Military Applications, "Clarification of NTS Off-site Criteria," memorandum, February 21, 1957). Hertford was confused as to what action the AEC had taken and pleaded that the new standard be made public so as to forestall a move to impose the more restrictive NAS standard (Hertford, "Clarification").

61. Strauss to Eisenhower, December 21, 1956; Elliott to Starbird, March 1, 1957.

62. Elliott to Starbird, November 1, 1956.

63. AEC, minutes of meeting, February 27, 1957; "Radiation Annex to Public Information and Education Plan for Operation Plumbbob," circulated to commission on February 15, 1957, defendant's exhibit no. 390.

64. Elliott to Starbird, April 24, 1957; AEC press release, Washington, D.C., May 13, 1957.

65. Starbird to Fields, "Fallout Difficulty Factor for Operation Plumbbob Shots," August 21, 1957; DOE, "Workshop," vol. 2, p. 143.

66. Charles R. Steadman, Jr., "The Boltzmann 'Hot Spot,'" Weather Service Nuclear Support Service, prepared for the Department of Energy, November 1982, pp. 23, 24; DOE, "Workshop," vol. 1, p. 161. Libby's idea eventually worked against the interests of the AEC and its successor agency. At the trial and in studies commissioned by the Department of Energy in the 1980s, the government attempted to denigrate the theory of hot spots, which challenged the validity of the lower average dose and exposure figures.

67. DOE, "Workshop," vol. 2, p. 57; Daily Activity Log for Lincoln Mine Zone, V. L. Ruckner, September 1957. The army monitors were poorly trained, were rotated too frequently, and were not adept at public relations. PHS monitors recalled the incident of army monitors who dressed up in their otherworldly decontamination suits and "scared hell" out of the local residents (DOE, "Workshop," vol. 2, pp. 126–127).

68. Helen Fallini and Martha Bordoli Laird, interviews with author, Twin Springs Ranch and Carson City, Nev., April 7, 8, 1983; Laird, TT, pp. 6–8; Joe Fallini, interview with author, Twin Springs Ranch, Nev., April 8, 1983. The Fallinis were highly regarded by test site officials before Martin's

death. In 1955 an army major visited the Fallinis and reported that they were "intelligent, sincere, and genuinely hospitable people" (Maj. Grant Kuhn to Joe B. Sanders, Deputy Manager, Las Vegas Field Office, March 25, 1955, JH, vol. 2, p. 2561).

69. Laird, TT, pp. 8–9, 17–18; Laird testimony, JH, vol. 3, pp. 15, 23.

70. Ibid., pp. 23–24.

71. Ibid., pp. 19, 31, 35, 39; Ambrose, *Eisenhower*, vol. 2, pp. 398–399; Fields, TT, p. 6555; Laird interview; Dunning to C. L. Dunham, Director, Division of Biology and Medicine, "Information on Bordoli Case," December 13, 1957.

72. *Tonopah Times-Bonanza*, October 26, 1956.

73. Ibid., October 26, November 16, 1956.

74. Ibid., June 14, 1957; testimony of Mahlon E. Gates, JH, vol. 3, p. 89. Although the two sectors north of the test site recorded the most events having fallout per sector, quantitatively the most fallout was recorded to the east.

75. *Tonopah Times-Bonanza*, May 24, 1957.

76. William P. Becko, interview with author, Tonopah, Nev., April 1983; Elliott to Salisbury, "Letter to Commissioner Libby," April 5, 1955.

77. Becko interview; Elliott to Crandall, April 5, 1955. Mrs. Crandall was a dogged reporter. She heard a radio interview of M. Stanley Livingston, a professor of physics at the Massachusetts Institute of Technology, who talked about fallout. She telephoned Livingston and complained about the lack of information given out by the AEC. Livingston wrote his friend Commissioner Libby and passed on the Crandall complaint. Libby ordered the AEC staff to investigate (Livingston to Libby, March 31, 1955; Libby to Livingston, May 9, 1955).

78. "Paul Jacobs and the Nuclear Gang," transcript of television documentary, WNET/13, 1979.

79. Shelby Thompson, Acting Director, Division of Information Services, to Starbird, "Meeting with Paul Jacobs, Correspondent for Reporter Magazine," March 1, 1957.

80. Thompson to all AEC commissioners, memo accompanying advance copy of the magazine, May 8, 1957.

81. Paul Jacobs, "Fallout From Nevada," *Reporter*, May 16, 1957. Thirty years later the story still reads well. Those writers who came later, and who had the benefit of the massive documentation that Jacobs lacked, could appreciate the hard digging and intuitive leaps that he made. Jacobs was out there first on these issues, and his work has held up exceedingly well over the years. A Salt Lake City journalist, referring to the claim of the *Deseret News* in 1977 that it was first with the story, wrote, "About the only thing the media can say is that we collectively did a pretty poor job. It took us

20 years to start asking follow up questions to Paul Jacobs' work" (letter to editor, Karl Idsvoog, *Deseret News*, December 29, 1978).

82. "Nuclear Gang," transcript; Elliott to Jacobs, March 10, 1958. Elliott said that the AEC wanted a record of what was asked and what was answered (Oliver R. Placak, Officer-in-Charge, PHS Off-Site Activities, Las Vegas, to W. W. Allaire, Director, Nevada Operations, "Interview with Paul Jacobs of Reporter Magazine," March 14, 1958). Placak enclosed a transcript of the interview and said that he had had previous contacts with Jacobs, which raises the interesting question of whether Jacobs got the PHS report from Placak's desk. (Paul Jacobs, "Precautions Are Being Taken by Those Who Know: An Inquiry Into the Power and Responsibilities of the AEC," *Atlantic*, February 1971, pp. 45–56; *Deseret News*, May 9, June 5, 1957; John A. Harris, Director, Division of Public Information, to commissioners, memorandum, August 25, 1970). Harris said that the general theme of the article would be "the low level of AEC credibility."

83. AEC, information meeting item, "Rebuttal to Paul Jacobs," February 4, 1971. Again, the AEC had advance notice of what would be said. It managed to get a copy of the script a day before the show was taped (Harris to commissioners, January 11, 1971).

84. Saul Landau, interview with author, Washington, D.C., March 17, 1983; Eve Pell, "Paul Jacobs: Death of an Investigative Reporter," *Politics and Other Human Interests*, February 14, 1978.

85. *Washington Post*, February 24, 1979.

86. Starbird to Reeves, "Smoky," June 17, 1957; Udall to Haralson, November 14, 1980. Udall said he got his information from two steelworkers who helped build the tower. Perhaps the AEC was seeking to experiment with coal rather than sand.

87. AEC press release, Project Smoky, August 27, 1957, CIC no. 34125.

88. Defense Nuclear Agency, "Shot Smoky," May 31, 1981; DOE, "Workshop," vol. 3, pp. 9, 13, 28; NOAA, "Analysis of Operation Plumbbob Nuclear Test Smoky Radiological and Meteorological Data," Weather Service Nuclear Support Office, prepared for the Department of Energy, September 1982, p. 10.

89. Starbird to all facilities, "Announcement on Plumbbob," September 26, 1957.

90. AEC, "Background Information on Nevada Nuclear Tests," Office of Test Information, Las Vegas, September 15, 1958.

91. Sanders to Elliott, "Blast Claim Damages," October 9, 1953; *Deseret News*, April 29, 1953, May 4, 1955; *Salt Lake News and Telegram*, May 2, 1955.

92. *Deseret News*, October 31, 1958; Divine, *Blowing on the Wind*, pp. 232–233. According to Divine's book, the fallout over Los Angeles at that

time was 120 times background readings and the highest ever recorded outside the immediate test site area (pp. 232–233).

93. DOE, "Workshop," vol. 2, p. 48.

94. M. D. Nordyke and M. N. Williamson, "Final Report, The Sedan Event," Lawrence Radiation Laboratory, August 6, 1965; O. R. Placak, "Final Off-Site Report, Project Sedan," Public Health Service, April 25, 1963; "Proceedings of Sedan Fallout and Radiochemistry Meeting," January 24, 1963, CIC no. 10745; memorandum from Richard L. Blanchard, in charge of biological experiments for Sedan, July 24, 1962; AEC, Project Manager's Report, "Plowshare Program, Project Sedan," prepared by the Nevada Operations Office, May 1963; Duncan Clark, Director, Division of Public Information, to AEC commissioners, "AEC Reports Additional Data on Sedan," draft press release, October 1, 1962, CIC no. 31653; John S. Kelly, Director, Division of Peaceful Nuclear Explosives, to A. R. Luedecke, AEC General Manager, "Preliminary Report on Iodine-131 from the Sedan Event," October 1, 1962; AEC, "Plowshare Program, Project Sedan," June 19, 1962, CIC no. 16829.

95. Gates, JH, vol. 3, p. 56.

96. Brig. Gen. Delmar L. Crowson, Director of Military Applications to AEC commissioners, "Pike Event," March 16, 1964. I came across mention of the Pike incident while reading the rich load of material contained in the 1980 transcript of the offsite monitors conference conducted by DOE in Las Vegas. I then followed up that revelation by searching the old AEC archives in Germantown, Md., and the DOE/CIC archives in Las Vegas. As far as I know, this is the first full disclosure of the Pike incident and its cover-up.

97. *Las Vegas Review-Journal*, March 15, 1964.

98. James E. Reeves, Manager, Nevada Operations Office, to Henry G. Vermillion, Director, Office of Information, "NTS Radiation Releases and Public Reporting," March 18, 1964, CIC no. 35826; "Approved Statement for Answer to Inquiry," March 14, 1964.

99. AEC, Nevada Operations Office, press release, March 16, 1964.

100. DOE, "Workshop," vol. 2, p. 203.

101. Reeves to Duncan Clark, "Proposed Public Report on Pike Venting," April 13, 1964; G. S. Douglas, for the files, "Request from Dunning Through Reeves and Phone Call From Jezik," June 29, 1964, CIC no. 35845; Douglas, for the files, July 1, 1964, CIC no. 35846; Placak, for the files, August 25, 1965.

102. Palfrey to Johnson, April 2, 1964. Bundy had no recollection of this incident (McGeorge Bundy to author, June 7, 1984).

103. Clark to commissioners, "Replies to Inquiries Relating Pike Venting to Nuclear Test Ban Treaty," June 3, 1964; Clark to commissioners, "PHS

Paper," December 24, 1964; Dunning to Crowson, "Fresh Fission Products in Air Filters at Los Angeles," April 23, 1964.
104. Testimony of Col. Raymond E. Brim, JH, vol. 3, pp. 265–266; 269–291.
105. *Deseret News*, December 18, 1968, February 29, 1969.
106. *New York Times*, October 11, 21, 1984.
107. Monitors' Reports, Baneberry, CIC no. 3103; AEC, "Baneberry Summary Report," May 1971; plaintiffs' and defendants' posttrial briefs, *Roberts and Nunamaker v. United States of America*; Larry C. Johns, interview with author, Las Vegas, Nev., April 1983.
108. Holifield to Seaborg, December 22, 1970; Edward B. Giller, AEC, Washington, to R. E. Miller, AEC, Las Vegas, December 22, 1970.
109. *Roberts and Nunamaker v. United States of America*, Partial Decision, Negligence Issue, June 8, 1982; Partial Decision, Radiation Dose and Causation Issue, January 20, 1983.

Chapter 7: The Victims

1. "A Brief History of the Tonopah Area," chamber of commerce leaflet; *San Jose Mercury-News*, August 29, 1982; "Tonopah: Long Live the Queen," *Sierra Life*, September/October 1982; *Tonopah Times-Bonanza*, January 9, 1953, April 7, 1983.
2. Bureau of Land Management, "Caliente Environmental Statement, Proposed Livestock Grazing Management Program," Las Vegas BLM District, 1979, pp. 2–72, 2–73.
3. Wesley Koyen, interview with author, Rachel, Nev., April 1983; Stanley W. Paher, *Nevada Ghost Towns and Mining Camps* (Berkeley: Howell-North Books, 1970), p. 301; Eva Hyde Koyen, *Treasures of Tempiute* (Sparks: Western Printing and Publishing Co., 1967), pp. vii, 94, 119. James W. Hulse, *The Nevada Adventure: A History* (Reno: University of Nevada Press, 1981), p. 199. Eva Koyen, who lived at the mine, wrote, "It was during the winter [of 1951] that we saw the first atomic bomb go off on Frenchman's Flat. What a weird light it is!"
4. Works Progress Administration, *Nevada*, American Guide Series, 1940, p. 177; James W. Hulse, *Lincoln County, Nevada, 1864–1909* (Reno: University of Nevada Press, 1971), pp. 13–14, 68; John M. Townley, *Conquered Provinces: Nevada Moves Southeast, 1864–1871* (Provo, Utah: Brigham Young University Press), pp. 13, 21, 30–31, 35.
5. Juanita Brooks, *The Mountain Meadows Massacre* (Norman: University of Oklahoma Press, 1962); Udall interview.
6. *Deseret News*, August 10, 1979; *People*, November 3, 1980. The con-

nection between the fallout and all the cancers of the cast and crew—91 cases of cancer among 220 people—is uncertain, especially in some cases where the type of cancer has no link or only a tenuous association with radiation. The Hollywood filmmakers were not represented in the Allen suit.

7. Nadine Henley to Libby, May 29, 1957, CIC no. 27408; Willard Libby to Henly, June 7, 1957; AEC press release, April 24, 1969; Michael Drosnin, *Citizen Hughes* (New York: Holt, Rinehart and Winston, 1985); Titus, *Bombs in the Backyard*, 99; *Las Vegas Review-Journal*, December 6, 1978.

8. Ralph Johnson, assistant U.S. attorney, interview with author, Salt Lake City, December 14, 1982; *Deseret News*, January 26, 1977, May 17, July 20, 1979, January 2, September 1, 1981, March 4, 1982. Meier had a checkered past. He had traveled with President Nixon's brother, Donald, to the Dominican Republic, after which the president ordered the phone of his younger brother tapped; he obtained Hughes's files in Mexico through the intercession of a member of the Canadian parliament; he was indicted for income tax evasion and skipped the country; he inflated the price of mining claims that he had acquired for Hughes; and he was wanted in California on a charge stemming from the stabbing death of a former business associate in a Beverly Hills hotel. In the year of the Allen trial, Meier was in a Canadian jail with a heart condition awaiting extradition to California. See *Deseret News* issues cited above.

9. Hazel Bradshaw, ed., *Under Dixie Sun* (St. George: Washington Chapter, Daughters of Utah Pioneers, 1950), p. 38. Although the copyright notice is dated 1950, the book has been updated to the late 1970s. Bureau of Land Management, "Environmental Assessment: Allen-Warner Valley Energy System," vol. 4, "Warner Valley Station," September 1977, pp. 3–6.

10. "The West as Victim," *High Country News*, October 31, 1983; Grant B. Harris, St. George, to Dale Haralson, April 16, 1979. Harris said he had been hired as an expert to judge mining claims. "In some localities where heavy concentrations of fallout were deposited, mining claims sold to gullible buyers for outrageous prices," he said. There were other accounts of this practice. It fell into a category of "salting" mining claims that has a number of precedents in the West.

11. Robinson, *History of Kane County*, p. 480.

12. "Report on Livestock Conditions Adjacent to the Las Vegas Bombing Range," L. A. Stoddart, range ecologist, Utah Agricultural Experiment Station, June 22, 1953, JH, vol. 2, pp. 1566–1568; statements of ranchers, dated August 25, September 9, October 29, 1953, JH, vol. 2, pp. 1586–1587; Stephen Brower, TT, pp. 2261–2262.

13. "Sheep Deaths in Utah and Nevada Following the 1953 Nuclear

Tests," prepared testimony of Harold A. Knapp, August 1, 1979, JH, vol. 1, p. 518; memorandum opinion, *Bulloch v. The United States of America*, Judge A. Sherman Christensen, United States District Court, August 4, 1982, pp. 17–18.

14. Testimony of Stephen Brower, JH, vol. 1, pp. 237, 240–241; "Statements of a Panel of Utah Citizens to Discuss the Impact of Radiation on Sheep," JH, vol. 1, pp. 227–243; Joe B. Sanders, Deputy Field Manager, Las Vegas Field Office, for the files, "Sheep Losses—Cedar City Area," undated, plaintiffs' exhibit no. 600.

15. F. H. Melvin to Chief, Bureau of Animal Industry, June 8, 1953, JH, vol. 3, pp. 715–716, 721; Monroe A. Holmes, veterinarian, PHS, to M.O.C., Communicable Disease Center, Atlanta, "Compiled Report on Cooperative Field Survey of Sheep Deaths in SW Utah," undated, JH, vol. 3, p. 721 (the undated copy covers activities from late May through June 1953); W. W. Allaire, Chief, Operations Branch, Office of Test Operations, Las Vegas, to files, "Report of Trip to Salt Lake City Meeting RE Nevada-Utah Livestock," undated, CIC no. 371 (the report was retyped in 1981 by the DOE from a poor photocopy, and the date was left off; it covers a meeting that took place on August 3–4, 1953).

16. Chief, Veterinary Public Health, to Executive Office, "Epidemic Aid Request from Dr. G. A. Spendlove, Health Officer, Utah," June 4, 1953, JH, vol. 2, pp. 1506–1507.

17. Testimony of Jack Pace, JH, vol. 1, p. 230; Brower testimony, TT, pp. 2219–2220.

18. Holmes, "Sheep Deaths," p. 733.

19. Wolff to William Hadlow, PHS, Rocky Mountain Laboratory, June 10, 1953, JH, vol. 1, p. 534; Veenstra to Allaire, June 17, 1953; JH, vol. 1, pp. 536–537; Thompsett to R. B. Cole, AEC Office of Engineering and Construction, "Possible Radiation of Animals," undated, JH, vol. 1, p. 864.

20. AEC, minutes of meeting, June 10, 1953, JH, vol. 1, p. 157; AEC, minutes of meeting, June 17, 1953, JH, vol. 1, p. 202; AEC, minutes of meeting, July 15, 1953, JH, vol. 1, p. 162; Bugher to Fields, "Damage to Horses from Beta Burns," memorandum, August 7, 1953, plaintiffs' exhibit no. 304; Bugher to M. W. Boyer, AEC General Manager, "Death of Cattle Adjacent to Test Site," August 20, 1953, EV, plaintiffs' exhibit no. 303; Joe B. Sanders, Acting Field Manager, Las Vegas Field Office, to Elliott, "Settlements on Horses and Blast Damage," October 9, 1953. The AEC settled the claims of the Stewart brothers of Alamo for damage to twenty horses for between $250 and $300 a horse. Sixteen of the irradiated horses were sold and slaughtered for chicken feed, and in such a manner the radiation passed through the food chain.

21. Selby Thompson, Acting Director, Division of Information Services,

to general manager and commissioners, AEC, "NPG Operating Statement," August 3, 1953.

22. Fields to Tyler, June 16, 1953.

23. Pearson to Bugher, "Livestock Losses Around Test Site," June 21, 1953, JH, vol. 2, pp. 1540–1541, 1549.

24. Brower testimony, TT, p. 2257. The three times Brower was under oath were at the 1979 congressional hearing, the second sheep trial, and the Allen trial. He repeated the charges in a letter to Governor Scott M. Matheson and in a separate letter to Matheson's aide, Mike Zimmerman. In the letter to the governor, Brower stated, "The original, preplanned strategy of the AEC worked because they had the indiscriminate power to intimidate, to withhold, screen, change, and classify any and all information, reports and data" (Brower to Matheson, "Radioactive Fallout and Sheep Death in Iron County, 1953," February 14, 1979, and Brower to Zimmerman, "Radioactive Fallout and Sheep Deaths in Iron County, 1953," February 16, 1979, JH, vol. 2, pp. 1849–1852; *Deseret News*, February 15, 1979).

25. Brower, TT, pp. 2238–2240. One of the provisions of the study was that it would not consider radiation as a cause of the sheep deaths. The results of the study were inconclusive (Project Outline, Utah State Agricultural College, 1954; Holmes to Wolff, May 24, 1954). Holmes, a PHS veterinarian, said the study was just an extension of a feeding and nutritional inquiry that had been under way for a number of years, and he had objected to its not being more specifically aimed at the recent sheep losses. The implication here is that the Utah academics were bought off, as they may have been in the 1980s when they successfully chased after millions of dollars in federal funds for human health studies with the help of Senator Orrin Hatch. Utah congressman Douglas R. Stringfellow pursued the funds for Utah State in 1954 (Stringfellow to Strauss, February 10, 1954; Strauss to Stringfellow, March 9, May 4, 1954, JH, vol. 2, pp. 2557–2559).

26. Brower testimony, JH, vol. 1, p. 233.

27. Pearson to Bugher, "Salt Lake City Conference on Livestock Losses," August 9, 1953; Allaire, "Nevada-Utah Livestock." Fortunately, there are two accounts of the meeting.

28. Notes taken on August 9, 1953, meeting, JH, vol. 2, pp. 1578–1585.

29. C. E. Lushbaugh et al., "Comparative Study of Experimentally Produced Beta Lesions and Skin Lesions in Utah Range Sheep," Los Alamos Scientific Laboratory, November 30, 1953, JH, vol. 2, pp. 1708–1730; L. K. Bustad et al., "A Comparative Study of Hanford and Utah Sheep," Biology Section, Radiological Sciences Department, November 30, 1953, JH, vol. 2, pp. 1731–1755; Warren E. Burger, Assistant Attorney General, Civil Division, to General Counsel, AEC, June 20, 1955. The Burger letter outlined the activities of John J. Finn, a Justice Department lawyer who collected

evidence for the 1956 sheep trial. Finn talked with Dr. Leo Bustad, a veterinarian at the Hanford laboratory in the state of Washington who had carried out the experiments and who pointed out that those experiments and similar ones at Los Alamos had been conducted on healthy sheep, not the undernourished animals that were in the Nevada desert that spring. This might cause problems at the trial, Finn thought. "Dr. Bustad admitted that there might be some merit in this contention and indicated he would immediately conduct some experiments on sheep which would be starved. Such experiments may have ultimate value in these cases," noted the letter (background for and exchange of correspondence between Harold A. Knapp and Dr. Leo Bustad, JH, vol. 3, pp. 957–964). Knapp, never one to pull his punches, looked backward in time and accused Bustad of suppressing evidence, a charge that Bustad, dean of the College of Veterinary Medicine at Washington State University in 1980, denied. The journal *Science* politely called it a "discrepancy" between reports ("Scientists Implicated in Atom Test Deception," *Science*, November 5, 1982). What was the dose to the sheep? Knapp estimated 3,000 to 5,000 rads to the gastrointestinal tract. The government said 100 to 400 rads (Knapp testimony, JH, vol. 3, p. 294; Lynn Anspaugh and John J. Koranda, eds., "Preliminary Assessment of Radiation Dose to Sheep Wintering in the Vicinity of the Nevada Test Site in 1953," draft revision 3, April 2, 1982). Anspaugh, of the Livermore laboratory, was a task group leader for the Offsite Radiation Exposure Review Project of DOE and testified for the government at the Bulloch II and Allen trials. This latest government report on the sheep deaths labeled the cause an enigma (Anspaugh and Koranda, "Preliminary Assessment").

30. Holmes to Medical Officer in Charge, CDC, memorandum, November 9, 1953, JH, vol. 3, pp. 701–704; Gordon Dunning, "Los Alamos Conference on Livestock Losses," October 27, 1953, JH, vol. 3, p. 778; Gordon Dunning, TT, p. 4407.

31. Holmes, November 9, 1953, JH, vol. 3, p. 703; Arthur H. Wolff, "Report of Trip to Los Alamos, October 26–29, 1953, JH, vol. 3, p. 700; Dunning, TT, pp. 4407–4409, 4469–4470, 4470–4472. At the Allen trial, Dunning testified, "There was no coverup. There was no misrepresentation that I could see." Nor did he recall making the purse strings comment. Dunning said he acted as a "scientific secretary" and got the panel members to summarize their discussion in a coherent statement (Dunning to author, "Comments" on manuscript, attachment 5, undated; see chapter 8, note 9, for an explanation of the "Comments").

32. Dunning to Salisbury, "Alleged Radiation Damage to Sheep," November 3, 1953.

33. "Sheep Losses Adjacent to the Nevada Proving Grounds," report by the director, Division of Biology and Medicine, November 4, 1953.

34. Pearson to Veenstra, October 19, 1953; Veenstra to Pearson, December 23, 1953; Holmes to Spendlove, "Observations and Comments on Draft of Paul Pearson's Final Report on Sheep Losses," December 16, 1953; Sanders to Trum, October 19, 1953; Sanders for files, undated.

35. AEC press release, January 13, 1954, EV, plaintiffs' exhibit no. 120. The AEC's fifteenth semiannual report also cited the Los Alamos and Hanford studies as demonstrating that radiation levels were too low to cause the sheep deaths, but no other causes were named (AEC, *Fifteenth Semi Annual Report of the Atomic Energy Commission*, July–December 1953, p. 51; W. T. Huffman, Bureau of Animal Industry, Salt Lake City, to H. W. Schoening, Director, Pathological Division, Bureau of Animal Industry, Department of Agriculture, June 23, 1953).

36. *New York Times*, January 17, 1954.

37. Meeting of livestockmen and AEC officials, Cedar City Fire House, January 13, 1954, JH, vol. 2, pp. 1794–1801.

38. Maj. Grant Kuhn for the files, "Trip Report of Major Grant Kuhn, March 9–22, 1955, JH, vol. 2, pp. 2560–2562.

39. Lt. Col. Bernard F. Trum, "Report of Lt. Col. Bernard F. Trum, March 31 to April 20, 1955," JH, vol. 2, pp. 2576–2599.

40. Trum to Veenstra, March 25, 1955, CIC no. 14241.

41. Trum report, March 31–April 20, 1953.

42. Veenstra to Trum, two letters, April 7, 1953. At the top of the typed letter was the notation: "This letter has not been sent, has not been published, and is confidential to the department." It is difficult to state exactly what pressure was brought to bear on Veenstra. Was he called on to demonstrate loyalty to the Army Veterinary Corps, to his government, or to what, and for what reason? The handwritten note was a "Dear Bernie" letter. It asked that his regards be given to Sanders of the AEC and cited the meagerness of the evidence supporting radiation and "all your data."

43. Dan Bushnell, *Bulloch*, JH, vol. 1, p. 1014; Trum to Thompsett, May 9, 1955; deposition of Thompsett, April 6, 1956, *Bulloch*, JH, vol. 1, pp. 850–863; Brower, TT, p. 2237. AEC employees and government lawyers would go out of their way to denigrate Thompsett. Sanders testified that he was drunk most of the time, and Henry Gill frequently disparaged him at the second sheep trial and the Allen trial (testimony of Joe Sanders, *Bulloch*, vol. 1, p. 1328–1329). Shortly before his death, Thompsett told a reporter in 1979 that he still believed that radioactive fallout killed the sheep (*Deseret News*, March 8, 1979).

44. Trum to Thomas, May 12, 1955.

45. I believe that this is the first time that Warren Burger's name has been associated with the cover-up. I came across his name, and his role in the sheep case, in the following manner. From a very brief description, I called for some documents from the CIC files in Las Vegas that seemed

to bear on the sheep case. When I got home, I read them. The name of
Warren E. Burger was at the end of two letters that had come from the
old AEC files. The current edition of *Who's Who* confirmed that Burger
had served as assistant attorney general from 1953 to 1956. I then asked
Henry Gill, the government's chief attorney at the second sheep trial and
the Allen trial, about Burger's role. Gill, a loyal government attorney and
former employee of the federal court system, referred me to the transcript
of the second sheep trial, where a "Warren Burgar" is briefly mentioned. I
had read the transcript before but had missed the fleeting reference. I then
wrote Burger, citing what the letters stated, and asked him to what extent
he had been knowledgeable about preparations for the trial and responsible
for conduct of the case. Burger's administrative assistant, Mark W. Cannon,
replied: "If your statement is correct that this case is now pending in a
Federal Court, it would be inappropriate for the Chief Justice to comment
on any of the points you have mentioned" (Cannon to author, August 21,
1984). Few people or institutions that were party to this sad episode in the
country's history came away untarnished. This is true from the chief justice,
to congressional leaders, to presidents of both parties, and on down.

46. Burger to General Counsel, "Re: David G. Bulloch et al. v. U.S.,"
June 20, 1955, CIC no. 25993 (two letters). There is no CIC number on
my copy of the second letter. Both letters seem to be retyped copies of the
originals. Under the salutation, "Yours very truly," Burger's name is typed.
It seems that both letters may have been signed for Burger by Bonnell
Phillips, chief of the Torts Section. It would be surprising, and a great lapse
in responsibility, if Burger had not at least been told of their contents.

47. Dan S. Bushnell, "The Sheep Deaths and the AEC Cover-up," JH,
vol. 1, p. 521; Burger/Phillips to Chalmers F. King, AEC general counsel,
December 9, 1955; Burger/Phillips to Llewellyn O. Thomas, assistant U.S.
attorney, undated but from approximately the same time.

48. Finn, when questioned by Judge Christensen, JH, vol. 1, p. 1050.

49. *Bulloch v. United States*, civ. no. C-19-55, United States District
Court, Utah, Central Division, 133 F. Supp. 885; *Bulloch*, October 26,
1956, 145 F. Supp. 824.

50. Edward Diamond, AEC Deputy General Counsel, to George Norris,
Jr., Committee Counsel, Joint Committee on Atomic Energy, November 27,
1956.

51. Testimony of Libassi, JH, vol. 1, p. 98; testimony of Fredrickson,
JH, vol. 1, p. 99.

52. House Subcommittee on Oversight and Investigations, *The Forgotten Guinea Pigs*, 96th Cong., 2d sess., August 1980, p. 37.

53. *Los Angeles Times*, June 30, 1981, February 14, 1982; *New York
Times*, August 14, 1982.

54. Memorandum, Findings of Fact and Conclusions of Law, *Bulloch* v.

United States of America, United States District Court, District of Utah, Central Division, Judge A. Sherman Christensen, August 4, 1982; *Los Angeles Times*, August 5, 1982; *Deseret News*, August 5, 1982; *New York Times*, August 5, 1982. In addition to Finn and King, Dan Bushnell, the plaintiffs' attorney, named Llewellyn Thomas, United States Attorney in Salt Lake City, and two other AEC attorneys, Charles F. Eason and Donald Fowler, as participating in the fraud. Fowler later became the general counsel for the California Institute of Technology, and Eason was the Washington representative for a firm involved in the disposal of radioactive waste. It was remarkable how well all those who were involved in the sheep case did after leaving the umbrella of the AEC. Like Judge Christensen and Gill, Bushnell stopped short of naming Phillips and Burger ("Memorandum in Support of Motion for Summary Ruling of Fraud Upon the Court," *Bulloch* v. *U.S.*, p. 54).

55. Brief for the Appellant, *Bulloch* v. *U.S.*, Court of Appeals for the Tenth Circuit, February 1, 1983, p. 12.

56. Transcript of *Bulloch et al.* v. *United States of America*, May 11, 1982, pp. 3–9. By this time Phillips was dead. Gill said that he had talked to Finn, who was retired and living in Miami. Finn said that he was not aware of any review of his work "at this level," according to Gill. Whether that is true remains to be seen; it may be that Burger and Phillips just chose not to tell underlings what they discussed. Bushnell thought the situation was "a bit awkward." He was quite right.

57. Jenkins, TT, p. 1224.

58. Deposition of LeOra Hafen, September 18, 1980; LeOra Hafen, TT, pp. 1759–1781; Dr. James R. Miller, TT, pp. 1675–1686.

59. Helen Nisson, TT, pp. 317–353; William Sandberg, TT, pp. 353–366; Wanda Stevens, TT, pp. 366–375; Darrell Nisson, DP, September 17, 1980, TT, pp. 381–499.

60. Rula D. Orton, TT, pp. 437–455; Mary Dawn Dalton, TT, pp. 456–463; Warren A. Dalton, TT, pp. 463–468; Jean H. Hendrickson, TT, pp. 469–475.

61. Vonda Y. McKinney, DP, September 24, 1980, August 4, 1982; Neil Cooley, DP, August 4, 1982; Vonda McKinney, TT, pp. 2157–2204; Vonda McKinney interview.

62. Lorna Bruhn, DP, September 18, 1980, July 21, 1982; Hall Loraine Woodbury, DP, July 21, 1982; Lorna Bruhn, TT, pp. 1092–1145; Beth Bruhn Hurst, TT, pp. 1145–1151; Lorna Bruhn, interview with author, St. George, Utah, May 2, 1983; Elizabeth Wright, "My Candle Burns at Both Its Ends," *Utah Holiday*, February 1982, p. 38; *Dixie Sun*, May 26, 1953.

63. Blaine Hart Johnson, TT, pp. 1782–1819.

64. VeRene Esplin Tait, TT, pp. 1227–1246; VeRene Tait interview; Patti Tait Heaton, TT, pp. 1246–1253; Joy Reusch Jordan, TT, pp. 1253–1258.

65. Florence Crabtree, TT, pp. 1152–1178; Joyce Crabtree Messer, TT, pp. 1178–1187; John Albert Crabtree, TT, pp. 1187–1195; Isaac A. Nelson, TT, pp. 1195–1201.

66. Mildred Bowler, TT, pp. 1820–1854.

67. Juanita Orton, TT, pp. 1446–1466.

68. Delsa Bradshaw, TT, pp. 1580–1602.

69. Harold Thompson, TT, pp. 1550–1579.

70. Paul W. Wood, TT, pp. 1603–1619.

71. Dwight Pectol, TT, pp. 1259–1270; Delores Smith Davis, TT, pp. 1271–1283; Darrell Neal Stephens, TT, pp. 1283–1285; Dr. LaVere Erickson, TT, pp. 2890–2898.

72. Louise Ouida Aicher, DP, September 22, 1980, July 27, 1982; Aicher, interview with author, Hiko, Nev., May 4, 1983; Keith W. Whipple, DP, October 28, 1980, July 27, 1982; Keith Whipple, interview, May 3, 1983; Jane Whipple Bradshaw, DP, September 18, 1980, July 27, 1982; Constance Rowene Nelson, DP, July 27, 1982; Jane Whipple Bradshaw, TT, pp. 1710–1758; Jane Whipple Bradshaw, interview with author, May 4, 1983; various medical records from Sansum Medical Clinic, Santa Barbara Cottage Hospital, and Valley View Medical Center in Cedar City.

73. Delwin Leroy Wilson, TT, pp. 1286–1302; Susan Wilson Lloyd, TT, pp. 1302–1308; Dr. Norman H. Fawson, TT, pp. 2435–2446, 2899–2921.

74. James N. Prince, TT, pp. 1202–1221.

75. Curtis M. Berry, TT, pp. 387–413; Charlotte Gleave, TT, pp. 413–428; Ora Mitchell, TT, pp. 428–430.

76. Sharon Hunt, TT, pp. 1855–1874.

77. Jacqueline Sanders, TT, pp. 1876–1902.

78. William Swapp, TT, pp. 1687–1709.

79. Jeffrey W. Bradshaw, DP, September 18, 1980; Jeffrey W. Bradshaw, TT, pp. 1468–1503; Verlynn Bradshaw, DP, September 18, 1980; Verlynn W. Bradshaw, TT, pp. 1503–1519.

80. Lionel Walker, TT, pp. 1628–1666; LaRue Walker, TT, pp. 1666–1671.

81. Norma Jean Pollitt, TT, pp. 1520–1549.

Chapter 8: The Scientists

1. Shields Warren, "The Origin of Cancer in Man," *Radiology*, November 1945, pp. 614–615; K. Z. Morgan, "Historical Sketch of Radiation-Protection Experience and Survey of Existing Problems," Oak Ridge National Laboratory, TID-388, issued March 12, 1951 (written in 1949), pp. 1–3; Robert S. Stone, "The Concept of a Maximum Permissible Exposure," *Radiology*, May 1952, pp. 640–643. I have attempted to cite literature that would have been easily accessible to AEC scientists. Dr. Shields Warren

was director of the AEC's Division of Biology and Medicine in the late 1940s and early 1950s. Dr. Stone's work was financed by the Manhattan Project, for whom he worked in a highly placed position, and by the AEC. *Radiology* was the journal of the Radiological Society of America. It was one of those professional peer-review journals so beloved of scientists. The article was derived from the Carmel Lecture delivered to the society in December 1951. The dates, as they relate to the timing of the Nevada tests, are important (James F. Holland and Emil Frei, *Cancer Medicine* [Philadelphia: Lea & Febiger, 1973], p. 93).

2. Stone, "Maximum Permissible Exposure"; Dr. J. Furth, "Recent Studies on the Etiology and Nature of Leukemia," Analytical Review, Biology Division, Oak Ridge National Laboratory, accepted for publication July 13, 1951; Richard G. Hewlett, "Nuclear Weapons Testing and Studies Related to Health Effects: An Historical Summary," Draft VI, Prepared for Director, National Institutes of Health, October 1980, p. 14.

3. Morgan, "Radiation-Protection Experience," p. 6.

4. Atomic Energy Commission, "Report on Project Gabriel," December 12, 1949; Atomic Energy Commission, "Project Gabriel," February 14, 1952. Project Gabriel became Project Sunshine in the fall of 1953—an ill-named effort to monitor strontium 90 on a worldwide basis (Hewlett, *Health Effects*, p. 49).

5. Harold A. Knapp, Jr., TT, pp. 2560, 2607, 2611; Harold A. Knapp, Jr., "Computation of the Radiation Dose Which Might Have Accrued Had the Nuclear Cloud from the 42.7 KT Simon Shot of April 25, 1953, Experienced a Rainout at a Distance of 120 Miles from the Nevada Test Site Similar to the Rainout Which Occurred 36 Hours After Detonation in the Vicinity of Troy, New York," October 6, 1982, EV, plaintiffs' exhibit no. 1017.

6. Howard L. Andrews, "Radioactive Fallout from Bomb Clouds," *Science*, September 9, 1955, pp. 453–456.

7. Haralson to author, March 7, 1983.

8. Udall interview; Gordon Dunning, DP, September 9, 1982, p. 99.

9. Closing argument, Henry Gill, TT, pp. 111–114. "I submit, your Honor, that Gordon Dunning is not an ogre," said Gill (Hewlett, interview with author, Germantown, Md., March 9, 1983; Dunning, telephone conversation with author, April, 1983). Dunning said, "I did not have any unique or privileged access to the Commissioners." (Gordon Dunning, "Comments," p. 2). My efforts to interview Dunning were not successful. First he said that on the advice of Henry Gill, the government attorney, he would not speak to me while the sheep case was being appealed (author to Dunning, February 18, 1983; telephone conversation, April 1983). Gill, who at first was eager to review an early draft of this book and later refused to do so, sent the manuscript to Dunning, who then changed his mind about

seeing me. Dunning, who lived south of Tucson in Green Valley, Ariz., wanted to meet at the DOE facility in Las Vegas. "I suggest that the initial discussions be off the record and then, if we agree, to go on the record. I would be prepared to devote a full day to the discussions and would wish to have a third party present," wrote Dunning (Dunning to author, June 7, 1984). An off-the-record interview would be of no use to me, and lawyers were an expense and a bother that I was not prepared to shoulder for a simple interview. I suggested that he send me his typed comments on the manuscript (author to Dunning, July 18, 1984). Back came a carefully typed and tabulated 60-page response, including supporting documents, titled "Comments," which I have used as one of a number of sources for this book. I made corrections in the book manuscript where they were warranted and retained the original wording where other sources outweighed Dunning's statements. My professional judgment was the final arbiter. I believe that by such means I have served the twin gods of accuracy and fairness.

10. Dunning, DP, pp. 4–16; Gordon Dunning, TT, pp. 4358–4398.

11. DOE, "Workshop," vol. 2, p. 84; Morgan Seal to Bruce Church, August 8, 1981; Dunning, TT, pp. 4410–4412.

12. Gordon Dunning, "Effects of Nuclear Weapons Testing," *Scientific Monthly*, December 1955, pp. 265–270; Gordon Dunning, "Protecting the Public During Weapons Testing at the Nevada Test Site," *Journal of the American Medical Association*, July 16, 1955.

13. *Time*, February 28, 1955; *Newsweek*, February 28, 1955; *Life*, February 28, 1955; *Life*, March 21, 1955; *U.S. News & World Report*, May 13, 1955, pp. 60–69; *McCall's*, January 1957; *Better Homes and Gardens*, May 1957; *Life*, June 10, 1957, pp. 24–29.

14. Representative Chet Holifield, Chairman of the Joint Committee on Atomic Energy, to Virginia Davis, June 3, 1957; Holifield to Strauss, May 21, 1957.

15. Joint Committee on Atomic Energy, "The Nature of Radioactive Fallout and Its Effects on Man," 85th Cong., 1st sess., 1957, pp. 170–257.

16. R. E. Hollingsworth to Fields, "Discussion of Fallout Problems," June 25, 1957.

17. Dunning provided an explanation for the difference: "One of the basic problems was that there was an understandably strong desire on the part of all to provide early estimates of the total accumulated dose that might be accrued. It became necessary, therefore, to use early exposure *rate* readings and any other available data and extrapolate forward in time by estimating such factors as rates of radiological decay of the fallout debris and the effects of shielding and weathering. What is not always recognized is these preliminary estimates were later updated after all of the data, which were accumulated over long time periods, were analyzed" (Dunning, "Com-

ments," pp. 5–6). Dunning said the recommendation to use shielding and weathering factors came from the 1953 Committee to Study the Nevada Proving Grounds, of which he was a member (ibid., p. 8).

18. Testimony of Gordon M. Dunning, "Fallout from Nuclear Tests at the Nevada Test Site," prepared for presentation to the Joint Committee on Atomic Energy, May 1959.

19. Knapp, TT, pp. 2452–2459; Institute for Defense Analysis, "Staff Members," January 1981; James V. Giles and A. Robert Smith, *An American Rape* (Washington: New Republic Book Co., 1975); Harold A. Knapp, "A Report to the Governor of Maryland: Request for Full Pardon for Three Citizens of Montgomery County Awaiting Execution in the Maryland Penitentiary," July 6, 1963.

20. Harold Knapp, interview with author, Germantown, Md., March 9, 1983; Knapp to "Tom," May 8, 1982; Harold A. Knapp, Jr., DP, July 27, 1982, pp. 51–54; *Fairfax Journal*, May 12, 1982. Stilwell, Deputy Under Secretary of Defense for Policy, headed the Pentagon's internal security program and instituted a program of screening nonintelligence employees and civilian contractors with polygraphs. In an attempt to stop press leaks, he ordered lie detector tests of twenty-two top Defense Department officials —including himself—who had participated in a key meeting.

21. Knapp to Haralson et al., October 9, 1982; Knapp to Udall, July 11, 1982; Knapp to Judge Jenkins, September 17, 1982; Knapp, TT, pp. 1223–1224, 2447.

22. R. Jeffrey Smith, "Atom Bomb Tests Leave Infamous Legacy," *Science*, October 15, 1982; Knapp, TT, pp. 2460–2463, 2467, 2549; Knapp to John T. Conway, Executive Director, Joint Committee on Atomic Energy, September 9, 1963, JH, vol. 2, pp. 2011–2017.

23. Knapp, interview; Knapp, DP, p. 62; Stuart C. Black, "Internal Dosimetry and Nuclear Weapons Tests: A Limited Historical Survey," Nuclear Radiation Assessment Division, Environmental Monitoring Systems Laboratory, February 28, 1983, p. 5. The Black study was done for the DOE. Dunning implied that he was helpful, or at least not a hindrance. "The reports were as available to Dr. Knapp as to me," said Dunning ("Comments," p. 17).

24. Knapp to Dunham, "Transmittal of Report on Radioiodine," September 13, 1962, JH, vol. 2, pp. 2018–2023.

25. Dunham to Knapp, "Draft Document: Average and Above Average Doses to the Thyroid of Children in the United States from Radioiodine from Nuclear Weapons Tests," October 24, 1962, JH, vol. 2, p. 1984; Knapp, TT, p. 2475.

26. Knapp to Dunham, June 27, 1963, JH, vol. 2, p. 1996.

27. John Gofman, TT, pp. 3408–3409, 3604–3605; report to the AEC, attachment A, "Background," for August 15, 1963, meeting.

28. Langham to Knapp, April 19, 1963.

29. Knapp, June 27, 1963, JH, vol. 2, pp. 1999, 2000.

30. Dunham to Knapp, April 10, 1963.

31. Knapp to Dunham, April 16, 1963.

32. Langham to Knapp, July 17, 1963.

33. AEC, "Background," August 15, 1963.

34. Knapp, JH, vol. 2, pp. 2000, 2001; Langham to Dunham, August 6, 1963; Dunham to Knapp, August 15, 1963; Dunham to A. R. Luedecke, AEC General Manager, "Proposed Publication of a Report Entitled I-131 in Fresh Milk and Human Thyroids Following a Single Deposition of Nuclear Test Fallout, by H. A. Knapp," memorandum, August 13, 1963.

35. H. A. Knapp, "Iodine-131 in Fresh Milk and Human Thyroids Following a Single Deposition of Nuclear Test Fallout," Fallout Studies Branch, Division of Biology and Medicine, AEC, TID-19266, June 1, 1963.

36. Dr. H. A. Knapp, "Iodine-131 in Fresh Milk and Human Thyroids Following a Single Deposition of Nuclear Test Fallout," *Nature*, May 9, 1964.

37. K. Z. Morgan, interview with author, Washington, D.C., March 16, 1983; Morgan, TT, p. 2878.

38. Morgan to Knapp, June 3, 1964. Eight years earlier Morgan had written AEC commissioner Libby, "I realize fully well the need for certain atomic weapons tests. To be sure, we must evaluate this need against possible risks to man. This letter is to point out that I am not certain the risks at present are negligible because of possible safety factors" (Morgan to Libby, November 5, 1956, CIC no. 31990).

39. Morgan to Knapp, August 15, 1964.

40. Morgan, TT, p. 2869.

41. Morgan, TT, pp. 2874–2875.

42. Tompkins to Commissioner Haworth, "Status Report on Current Activities of the Federal Radiation Council Working Group," September 25, 1962. Dunning stated, "I was a member of the Federal Radiation Council's working group and actively participated in the development of radiation protection guidelines for radioiodine" (Dunning, "Comments," p. 15; Dunning, TT, p. 4396). The council was set up, after President Eisenhower signed the necessary legislation, to head off a move to take away the AEC's authority to set radiation standards and give it to the PHS. The council consisted of the heads of the AEC and the Department of Health, Education, and Welfare, plus three other federal agencies. The working group consisted of their designees, and as its name suggests, it did all the work. I. F. Stone wrote,

"The Federal Radiation Council soon proved to be no more than a facade behind which the military and the AEC continued to run the show" ("How the Pentagon Protects Itself from Fallout," *I. F. Stone's Bi-Weekly*, June 24, 1963, p. 1.

43. Morgan, TT, p. 3321. Morgan said he was particularly scornful— *ashamed* was the term he used—of the National Academy of Science's report on the death of a Japanese fisherman following the 1954 Bravo test which was blamed on hepatitis from blood transfusions when "it was a clear case of radiation syndrome," and the controversial 1979 report on low levels of radiation, known as BEIR III. Of the National Council on Radiation Protection and the International Council on Radiation Protection, of which Morgan had been an honored member for years, the retired government scientist said they had done a great deal of good and a great deal of harm. Since members nominated their successors, the chances were "essentially zero" that a dissenting viewpoint would be represented (Ibid., pp. 3252–3253, 3283–3284). That the National Academy of Science was unduly influenced by the AEC may be due, at least in part, to the fact that Charles Dunham, the director of the AEC's Division of Medicine and Biology during the Knapp-Dunning fracas, moved over to the NAS to become its medical director and appointed members of the BEIR committee (Gofman, TT, p. 3504). Philip Handler, president of the National Academy of Sciences, cited the "troubled history" of BEIR III in the transmission of the final report to the administrator of the Environmental Protection Agency. This report was flawed in a number of ways. The resulting 524-page book, published by the National Academy Press, has a good number of its pages printed upside down (Committee on the Biological Effects of Ionizing Radiation, *The Effects on Populations of Exposures to Low Levels of Ionizing Radiation: 1980* [Washington, D.C.: National Academy Press, 1980]).

44. *Washington Post*, August 17, 1963; Joint Committee on Atomic Energy, Subcommittee on Research, Development, and Radiation, "Fallout, Radiation Standards, and Countermeasures," 88th Cong., 1st sess., August 20, 21, 22, 27, 1963, p. 556; Richard Daly and Lindsay Mattison, "Nevada Fallout: Past and Present Hazards," *Bulletin of the Atomic Scientists*, April 1964, p. 556; Dunning to Luedecke, "Radioactive Fallout from Current U.S. Nuclear Tests," memo, May 22, 1962.

45. *Washington Post*, August 21, 1963; *New York Times*, August 22, 1963.

46. Arthur Tamplin, TT, p. 2277.

47. AEC press release, "New Studies Planned at AEC's Livermore Laboratory," May 31, 1963; Gofman, TT, pp. 3411, 3432, 3653.

48. Tamplin, TT, pp. 2294, 2299, 2300, 2306; Gofman and Tamplin to Dunham, "The Problem of Potential Thyroid Pathology in Washington

County, Utah," December 6, 1965, p. 1; May to Seaborg, November 29, 1963, JH, vol. 2, p. 2120.

49. Gofman and Tamplin to Dunham, "Potential Thyroid Pathology," p. 3; Dunning to D. A. Ink, Assistant General Manager, "Discussion with Drs. Gofman and Tamplin re. Fallout Studies," December 15, 1965. Dunning said that he had talked to Gofman and Tamplin only briefly and that he had suggested that yet another study be done "properly" with a control group (Dunning, "Comments," p. 22).

50. Arthur R. Tamplin and H. Leonard Fisher, *Estimation of Dosage to Thyroids of Children in the U.S. From Nuclear Tests Conducted in Nevada During 1952 Through 1957*, Lawrence Radiation Laboratory, UCRL-14707, May 10, 1966.

51. Dr. Alfonse T. Masi, "Field Trip to Evaluate Epidemiological Studies Regarding Possible Human Effects of Nuclear Weapons Testing in Nevada," 1964. Dr. Masi of the Johns Hopkins University School of Hygiene and Public Health, a consultant to the PHS on the thyroid and leukemia studies, sounded a warning note that was not heeded. Of Dunning's work he said, "The serious shortcoming in any dosage argument is that not all radiation exposures have been documented. In fact, the information on human radiation exposure is sketchy and in need of detailed reconstruction." Dunning maintains that these were not his estimates but were instead the work of the 1959 Test Manager's Committee to Establish Fallout Doses to Communities Near the Nevada Test Site, otherwise known as the Shelton committee (Dunning, "Comments," pp. 23, 26). This is hairsplitting. Dunning presented the data to a congressional committee, published the results under his name, devised the models on which the estimates were based, and was the chief technician whose results were rubber-stamped by the committee members, who worked for the AEC and the PHS. The dose estimates were known to Dr. Masi, other contemporaries, and DOE scientists as Dunning's work.

52. Heath to Chief, Epidemiology Branch, "Investigation of Leukemia, Washington and Iron Counties, Utah," July 5, 1961; Edward S. Weiss, Chief, Biometrics Group, to C. D. Carlyle Thompson, Director, Utah State Department of Health, October 22, 1964. Weiss enclosed a publication and tables that he had constructed showing that "Utah has the lowest overall rate in the nation" (Edward S. Weiss, "Leukemia Mortality in Southwestern Utah, 1950–1964," July 23, 1965, JH, vol. 2, pp. 2191–2216). Heath testified at the Allen trial that he did not know if the overall Utah cancer rate or a rate for a Mormon population that was not exposed to fallout was available at that time. Obviously his colleague, Weiss, possessed such information. "In any case, they used what was related to white populations in the U.S. as the basic comparison," said Heath (Clark W. Heath, Jr., TT, pp. 574, 585, 593).

This was a grave statistical mistake. Studies would show the Utah cancer rate was from 22 to 55 percent below the national average (see below, note 93). I tend to view this omission as incompetence rather than insidiousness, but one never knows.

53. Clark W. Heath, Jr., Chief, Leukemia Unit, Epidemiology Branch, to Chief, Communicable Disease Center, "Leukemia in Fredonia, Arizona," August 4, 1966; Heath, TT, pp. 524–529, 562, 595, 596, 597.

54. Heath to Director, National Communicable Disease Center, "Community Surveys in Monticello, Parowan and Paragonah, Utah," November 14, 1967; Heath to Director, National Communicable Disease Center, "Leukemia in Parowan and Paragonah, Utah," April 26, 1967; Peter McPhedran, Leukemia Section, Epidemiology Program, to Chief, Communicable Disease Center, "Leukemia in Monticello, Utah," July 5, 1967; Heath, TT, p. 595.

55. Warren Winkelstein, Jr., and Stephen B. Hulley, eds., "Report of the Panel of Experts on the Archive of PHS Documents: Effects of Nuclear Weapons Testing on Health," Department of Health, Education, and Welfare, vol. 1, 1979, pp. 110–111. The fox-in-the-chicken-coop syndrome may have been applicable here, as the same agency once again investigated its prior activities.

56. Winkelstein and Hulley, "Effects of Nuclear Weapons Testing," p. 101.

57. Oliver R. Placak, Officer in Charge, Southwestern Radiological Health Laboratory, to Thompson, Dunning et al., "Protocol for Environmental Radioiodine Study," May 10, 1963, JH, vol. 2, pp. 2625–2626.

58. Weiss to Dunning, June 14, 1965, with attachments; Dunning, draft letter used for basis of telephone conversation with Weiss, June 14, 1965.

59. Dunning to O. S. Hiestand, Jr., Assistant General Counsel, Operations, "PHS Studies in Utah," September 3, 1965, with six attachments; Dunning to D. A. Ink, Assistant General Manager, "PHS Studies in Utah," August 27, 1965, JH, vol. 2, pp. 2250–2251; Dunning to Ink, "Notes for your use at Commission Information Meeting regarding PHS studies in Utah and Mr. Weiss' article on similar subject," August 27, 1965; Dunning to Ink, December 16, 1965, JH, vol. 2, pp. 2252–2254.

60. Dunning to Hiestand, September 3, 1965, attachments 4 and 5; Dunning to Ink, August 27, 1965; Gordon M. Dunning, "Criteria for Establishing Short-Term Permissible Ingestion of Fallout Material," *American Industrial Hygiene Association Journal*, April 1958, pp. 111–120; Gordon M. Dunning, "Radioactivity in the Diet," *Journal of the American Dietetic Association*, January 1963, pp. 17–27. In the 1958 article Dunning wrote that "the adult thyroid is relatively insensitive to radiation" but that there is the possibility that tumors might be produced in children "at relatively low radiation exposures." Dunning cited low dose figures in the 1963 article.

The next year he admitted a grave error: "There can be circumstances where levels of iodine-131 in milk can be a more controlling factor than external gamma exposures that have hitherto been considered of prime interest for local fallout" (Gordon M. Dunning, *Health Aspects of Nuclear Weapons Testing,* AEC pamphlet, 1964).

61. Judson Hardy, for the record, "Meeting of PHS-AEC representatives: re. Utah study," September 2, 1965, JH, vol. 2, p. 225; *Deseret News,* January 29, 1979.

62. Hardy, "Meeting of PHS-AEC representatives."

63. Dunning to Ink, "PHS Announcement on Study of Possible Fallout Effects in Utah," August 31, 1965.

64. *Washington Post,* April 14, 1979, and various documents, JH, vol. 2, pp. 2256–2269.

65. Testimony of F. Peter Libassi, JH, vol. 1, p. 221.

66. Ink to commissioners, "USPS Epidemiology Studies in Southwestern Utah," September 9, 1965, JH, vol. 2, pp. 2221, 2222; Ink to Terry, September 10, 1965, JH, vol. 2, pp. 2243–2244.

67. Donald R. Chadwick, Chief, Division of Radiological Health, to Surgeon General, "Utah-Nevada Population Study," October 11, 1965, JH, vol. 2, pp. 2117–2119; *Deseret News,* December 2, 1965; Public Health Service, press release, October 28, 1965.

68. Hardy to Thompson, August 27, 1965; Chadwick to surgeon general, "Thyroid Abnormalities Revealed by Field Study of Utah and Arizona Students," October 4, 1965, JH, vol. 2, pp. 2114–2116.

69. *Deseret News,* October 29, 1965. A series of articles published that fall by Hal Knight, *Deseret News* science writer, dealt with fallout and cancer. It was a cautious series that relied heavily on state health officials, who did not want to rock the boat and upset the state's economy. One story on iodine 131 in milk was headlined "Utah Fallout Effect Nil" (*Deseret News,* November 23, 1965).

70. Judson Hardy to the Assistant to the Surgeon General for Information, "*Saturday Evening Post* Projected Story on Utah Fallout Studies," December 2, 1965.

71. Transcript, "Statistical Considerations of Field Studies on Thyroid Disease in School Children in Utah-Arizona," Twinbrook Research Building, Rockville, Md., December 3, 1965, JH, vol. 2, pp. 2123–2176.

72. Public Health Service, press release, March 16, 1966; *Deseret News,* March 16, 1966, February 3, 1967; *Time,* March 25, 1966.

73. "Joint Meeting of AEC, Division of Biology and Medicine, Division of Operational Safety, and USPS, Division of Radiological Health," Twinbrook Research Building, Rockville, Md., May 3, 1966, JH, vol. 2, pp. 2772–2775. Arthur Tamplin, who was working under Gofman on another radioiodine-thyroid study, was present at the meeting.

74. Marvin L. Rallison et al., "Thyroid Disease in Children," *American Journal of Medicine*, April 1974, pp. 457–463. Edward S. Weiss, "Thyroid Nodularity in Southwestern Utah School Children Exposed to Fallout Radiation," *American Journal of Public Health*, February 1971, pp. 241–249. A preliminary report by Weiss did find an increase in thyroid cancer but noted that further study was needed to bear out those results (Edward S. Weiss, "Surgically Treated Thyroid Disease Among Young People in Utah, 1948–1962," *American Journal of Public Health*, October 1967), pp. 1807–1814.

75. Marvin L. Rallison to Paul Ensign, March 12, 1970.

76. Rallison to Edy Thompkins, Office of Research, Environmental Protection Agency, February 11, 1972, JH, vol. 2, pp. 2853–2854.

77. Bibb to Jerry Fain, October 15, 1971.

78. Daniel A. Hoffman et al., "A Feasibility Study of the Biological Effects of Fallout on People in Utah, Nevada, and Arizona," Public Health Service, February 1981.

79. Jamie Kalven, "Atomic Veterans: The Legal Quandry," *Bulletin of the Atomic Scientists*, January 1983, pp. 26–30; Allan Favish, "Radiation Injury and the Atomic Veteran: Shifting the Burden of Proof on Factual Causation," *Hastings Law Review*, March 1981, pp. 933–974; Patrick Huyghe and David Konigsberg, "Grim Legacy of Nuclear Testing," *New York Times Magazine*, April 22, 1979; Defense Nuclear Agency, "Nuclear Test Personnel Review," fact sheet, October 1, 1979; "Section in Military Bill Limits Suits in Atomic Tests," *New York Times*, November 4, 1984. In addition, there are three books on the atomic veterans: Howard L. Rosenberg, *Atomic Soldiers: American Victims of Nuclear Experiments* (Boston: Beacon Press, 1980); Michael Uhl and Tod Ensign, *GI Guinea Pigs: How the Pentagon Exposed Our Troops to Dangers More Deadly Than War; Agent Orange and Atomic Energy* (Playboy Press, 1980); Thomas H. Saffer and Orville E. Kelly, *Countdown Zero* (New York: G. P. Putnam's Sons, 1982).

80. Testimony of Glyn G. Caldwell, TT, pp. 2093, 2096, 2111, 2132; Glyn G. Caldwell et al., "Leukemia Among Participants in Military Maneuvers at a Nuclear Bomb Test," *Journal of the American Medical Association*, October 3, 1980, pp. 1575–1577. Dr. Heath supervised the Caldwell study and was a coauthor of the *JAMA* article. Dale Haralson questioned Caldwell at the Allen trial. He asked about a study that showed that persons with cancer of the larynx, from which Haralson was suffering, were apt to contract leukemia after radiotherapy. Caldwell said there was an increased risk (TT, pp. 2094–2095).

81. Charles W. Mays, interview with author, Salt Lake City, Utah, May 10, 1983; *Salt Lake Tribune*, July 11, 1982; "Obituary, Robert C. Pendleton," *Health Physics Journal*, December 1982, p. xi.

82. *Salt Lake Tribune*, December 12, 1968; Utah State Department of

Health Activity Report, report of telephone conversation between Lynn M. Thatcher and Jack McBride, December 12, 1968, JH, vol. 2, p. 2779. The following year Pendleton told the Rocky Mountain Chapter of the Society of Nuclear Medicine, "In essence, the people of Utah and the Mountain West are being asked to bear the brunt of the hazards of the testing program" (*Deseret News*, April 17, 1969). Statements like this made his more cautious and quiet colleagues shudder.

83. *Deseret News*, January 27, 1979; "Scientist Renews Pleas for Radiation Study," University of Utah press release, June 8, 1978; Mays, interview; Pendleton deposition in *David L. Timothy* v. *The United States of America*. The Timothy suit involved Utah residents with cancer who lived in the foothills of the Uinta Mountains, where Pendleton measured relatively high amounts of fallout. The mountains, where a rainout was more apt to occur, were east of Salt Lake City and in the prevailing track of fallout.

84. Pendleton, deposition, *Timothy* v. *U.S.A.*; Robert C. Pendleton, "Iodine 131 in Utah During July and August 1962," *Science*, August 16, 1963; testimony of Charles W. Mays, Joint Committee, 1963, pp. 540–542.

85. Mays testimony, Joint Committee, pp. 544–545; Thompson to Pendleton, August 4, 1962, CIC no. 32692; *Deseret News*, August 22, December 19, 1963; Pendleton, deposition, *Timothy* v. *U.S.A.*

86. Dunning to N. H. Woodruff, Director, AEC Division of Operational Safety, "Dr. Pendleton's Paper 'Iodine-131 in Utah During July and August 1962,'" June 14, 1963.

87. Pendleton to Knapp, July 22, 1963.

88. Pendleton to Knapp, July 12, 1963.

89. Joseph L. Lyon, TT, pp. 610–612, 624; Lyon, interview with author, Salt Lake City, April 12, 1983.

90. Lyon, TT, pp. 625–626.

91. Ibid., pp. 626–627.

92. Ibid., p. 627.

93. Joseph L. Lyon et al., "Cancer Incidence in Mormons and Non-Mormons in Utah, 1966–1970," *New England Journal of Medicine*, January 15, 1976, pp. 129–132.

94. Ibid., p. 630.

95. Ibid., p. 631.

96. Ibid., p. 632.

97. Ibid., p. 707. Lyon said that he understood that the briefing in the Forrestal Building, where DOE headquarters was located, would be attended only by scientists. A crew from a Salt Lake City television station was turned away, but after the meeting Gordon Eliot White of the *Deseret News* came up and introduced himself to the astonished Lyon. White said that he had taped the interview. There was some discussion between university

officials, who did not want to jeopardize publication of the study, and the newspaper's editor, with the end result that the story was not published at that time (Lyon interview).

98. Lyon, TT, pp. 695–697, 736.

99. Minutes of first meeting of Fallout Steering Committee, December 14, 1978.

100. Caldwell, conference and call record, with John C. Villforth, Director, BRH, December 5, 1978, JH, vol. 2, pp. 2857–2858; Caldwell, Chief, Cancer Branch, for the record, December 22, 1978.

101. Lynn Lyon to Jim Reading et al., "Follow-up of an Exposed Cohort," November 6, 1978.

102. Joseph F. Fraumeni, Jr., Chief, Epidemiology Branch, NCI, to Special Assistant to Director, NIH, "Review of November 6, 1978 memo from Dr. Lynn Lyon," January 24, 1979; Gilbert W. Beebe, Expert, Clinical Epidemiology Branch, NCI, to Director, NCI, January 26, 1979.

103. Donald S. Fredrickson, Director, National Institutes of Health, to Governor Matheson, November 23, 1979 (this letter is a masterpiece in the art of saying you lost without saying you lost). Alene Bently, press secretary to Governor Matheson, "Summary of Events Leading to Health Effects Studies of Low Level Radiation," September 30, 1982; Chase N. Peterson, "A Chronologic Summary: Proposals to Study the Health Effects of Radiation in Southern Utah," December 14, 1981; Mike Zimmerman to Governor Matheson, "Radiation Studies," memorandum, November 20, 1979; Lyon, interview.

104. Special Study Section, Radiation Research Subcommittee, "Fallout and Human Effects from US Atomic Weapons Tests," July 1, 1980. After reading the negative review of the proposal, Harold Knapp noted, "I would guess that more information will come out of the trial of the forthcoming lawsuit than will come from the university trying to get a huge, unenthusiastic federal bureaucracy to fund a huge study" (undated handwritten note).

105. Joseph L. Lyon et al., "Childhood Leukemias Associated with Fallout From Nuclear Testing," *New England Journal of Medicine*, February 22, 1979, pp. 397–402; *Deseret News*, February 13, 1979; Utah Cancer Registry, "Cancer in Utah," Report No. 3, 1967–1977, September 1979; Joseph L. Lyon et al., "Further Information on the Association of Childhood Leukemias with Atomic Fallout," *Banebury Report 4*, 1980, pp. 146–162.

106. Lyon, TT, pp. 645–646; Lyon, JH, vol. 1, p. 355.

107. Lyon, TT, p. 728; Lyon, JH, vol. 1, p. 375.

108. Lyon, TT, p. 729.

109. Charles E. Land, TT, pp. 5080–5086.

110. C. E. Land, "The Hazards of Fallout or of Epidemiologic Re-

search?" *New England Journal of Medicine*, February 22, 1979, pp. 431–432.

111. Land, TT, pp. 5088–5090.

112. Herman Chernoff, "Report for the United States Accounting Office," 1979, p. 30.

113. Peterson, JH, vol. 1, p. 363.

114. Department of Health and Human Services, press release, September 22, 1980.

115. Peterson to Ron Preston and David N. Sundwall, staff members of the Senate Committee on Labor and Human Resources, December 16, 1981.

116. Alene Bently to Governor Matheson, "re: Radiation Studies," May 7, 1981; Lynn Lyon to Jim Reading et al., "Follow-up of an Exposed Cohort," November 6, 1978.

117. Peterson to O. R. Lunt, Director, Laboratory of Biomedical and Environmental Sciences, Los Angeles, September 24, 1981.

118. Peterson, December 16, 1981. A note Peterson wrote on a copy of the Lunt letter stated that a formal written request was received in December 1981, but it had been sent to the wrong address. Peterson also emphasized the wrong-address aspect in an interview (Peterson, interview with author, May 17, 1983). The principal investigator for UCLA was James E. Enstrom of the School of Public Health, who had evidently heard that I was writing a book that touched on this subject and sent me material that was critical of Lyon. Backbiting among scientists was prevalent (Enstrom to author, April 19, 1983; Enstrom to editor, *New England Journal of Medicine*, June 28, 1977, p. 1491).

119. Lyon, interview.

120. Maggi Wilde to Alene E. Bently, Governor Matheson's press secretary, August 5, 1981.

121. Schweiker to Hatch, May 10, 1982; *Salt Lake Tribune*, May 11, 1982; University of Utah Medical Center, news release, July 15, 1982.

122. Charles E. Land et al., "Childhood Leukemia and Fallout from the Nevada Nuclear Tests," *Science*, January 13, 1984, pp. 139–144; Carl J. Johnson, "Cancer Incidence in an Area of Radioactive Fallout Downwind From the Nevada Test Site," *Journal of the American Medical Association*, January 13, 1984, pp. 230–236. The Johnson study, done for the plaintiffs, went through an unusual three cycles of peer review. "I had to be absolutely sure it was responsible and valid because it flies in the face of Government policy and has such a huge potential economic impact," said George Lundberg, editor of the journal (*New York Times*, January 13, 1984).

123. Closing arguments, TT, p. 4.

124. Ibid., p. 64.
125. Ibid., pp. 88–89.
126. Ibid., p. 152.

Chapter 9: The Judgments

1. Judge Bruce S. Jenkins, *Irene Allen et al. v. The United States of America*, Memorandum Opinion, May 10, 1984; *New York Times*, May 11, 1984. The *Times* story on the judgment appeared on the front page.

2. *San Francisco Chronicle*, May 11, 1984; *Deseret News*, May 11, 17, 1984; *New York Times*, May 12, 13, 1984.

3. Jenkins, Allen case opinion, pp. 236–237.

4. Ibid., pp. 155–156.

5. Ibid., pp. 370–371.

6. Ibid., p. 261.

7. Ibid., pp. 270–271, 313–316.

8. Ibid., p. 317.

9. Ibid., p. 9. Actually, a number of plaintiffs also lived in the Colorado River Basin.

10. Ibid., p. 344.

11. Ibid., pp. 375–407.

12. Senate Committee on Labor and Human Resources, *Examining How Liability Should Be Assessed for Damages Caused by Low-Level Radiation Effects Which Appear as Cancer Years After Exposure*, 98th Congress, 2nd sess., September 18, 1984, p. 84.

13. *David Bulloch v. The United States of America*, Appeal from the United States District Court for the District of Utah Central Division, United States Court of Appeals, Tenth Circuit, Slip Opinion, November 23, 1983; *David Bulloch et al. v. The United States of America*, Plaintiffs-Appellees' Petition for Rehearing with Suggest for Rehearing in Banc, December 22, 1983.

14. *Deseret News*, January 22, 1986.

15. *Irene H. Allen et al. v. The United States of America*, Appeal from the United States District Court for the District of Utah, United States Court of Appeals, Tenth Circuit, published opinion, April 20, 1987, pp. 5–6, 9, 14.

16. Ibid., p. 14. At the congressional level, the probability tables requested by Senator Hatch's amendment were delivered by government scientists but proved to be too controversial. A Hatch amendment in 1985 that would have used the tables to help a commission hand out $150 million in compensation to downwind cancer victims was defeated in the Senate. The amendment, which was opposed by the Reagan administration, was

attached to a bill that passed. That bill provided an equivalent amount in compensation to the Marshall Islanders (*Congressional Record*, Senate, November 14, 1985, pp. 15582–15599). Hatch did manage to pass an amendment that authorized $4 million for a research center at the University of Utah to study "the health effects of nuclear energy and other new energy technologies" and $6 million for a cancer screening clinic and research center in St. George. The Dixie Medical Center Regional Cancer Center was dedicated in May 1987 (Labor and Human Resources Committee, news releases, October 9, 30, 1984; *Deseret News*, October 31, 1984; "Your Health Matters," Dixie Medical Center, May 1987). At the administrative level, DOE continued its Offsite Radiation Exposure Review Project, but the review group of scientists that had been created to give it credibility went out of existence in May 1987 with a final report that stated that the doses were, in fact, about the same that Gordon Dunning had reported in 1959. Chairman Edward L. Alpen of the Dose Assessment Advisory Group said, "But I guess the theme is, there were no surprises" (Fourteenth Dose Assessment Advisory Group meeting, May 21, 1987, transcript, vol. 2, pp. 19, 35).

17. *Allen* v. *United States*, appeal opinion, p. 15.

18. *New York Times*, January 12, 1988. So much for the short life of "a landmark ruling." The *Times* story was buried on an inside page in a wrap-up of other Supreme Court actions. The media had gone on to other concerns.

INDEX

ABOUT THE AUTHOR

PHILIP L. FRADKIN grew up in
New Jersey and graduated from Williams College in 1957. He has been a reporter for various California newspapers, including the *Los Angeles Times*,
where he shared a Pulitzer Prize for the newspaper's coverage of the Watts
riot, covered the Vietnam War, and was an environmental writer. He was
assistant secretary of the California Resources Agency during the administration of Governor Edmund G. Brown, Jr., and later served as western
editor of *Audubon* magazine. The author of four books, Fradkin has taught
writing at Stanford University and the University of California at Berkeley
and is the editor and publisher of Redwood Press, a small book publishing firm in Marin County, California. He is married and the father of two
children.